An Introduction to
Creating Standardized Clinical Trial Data with SAS®

Todd Case
YuTing Tian

sas.com/books

Contents

About This Book

What Does This Book Cover?

The purpose of this book is to introduce standardized clinical trial data to anyone interested in understanding the pharmaceutical industry and how that data is collected and created.

This book introduces the concept of standardized clinical data, technical terms, and programming practices in the pharmaceutical industry as well as clear and concise explanations with numerous practical examples. We include basic knowledge of the pharmaceutical industry as well as SAS programming practices used in the industry.

This book does not cover how to create define.xml, although we do introduce it to the reader.

What Are the Prerequisites for This Book?

The only prerequisite for this book is an interest in the pharmaceutical industry.

What Should You Know about the Examples?

This book includes SAS code and simulated data for the reader to gain hands-on experience with standardized clinical data. Visit the author's page at http://support.sas.com/case to access the example code and data.

Software Used to Develop the Book's Content

SAS Version 9.4 was used to develop the content and examples in this book.

Example Code and Data

This book includes data and complete programs used to create simulated standardized clinical trial data. Visit http://support.sas.com/case to access the example code and data.

An example to derive sex in the Demographics domain is demonstrated below:

```
/*Derive SEX*/
if SEX_="Female" then SEX="F";
else if SEX_="Male" then SEX="M";
else if SEX_="Unknown" then SEX="U";
else if SEX_="Undifferentiated" then SEX="UNDIFFERENTIATED";
```

SAS OnDemand for Academics

This book is compatible with SAS OnDemand for Academics. If you are using SAS OnDemand for Academics, then begin here: https://www.sas.com/en_us/software/on-demand-for-academics.html.

Acknowledgments

Thank you to CDISC and the technical reviewers who provided feedback: Margaret Hung, Matt Becker, Peter Eberhardt, Laura Elliott, William Kuan, and Crystal Cheng.

We Want to Hear from You

SAS Press books are written *by* SAS Users *for* SAS Users. We welcome your participation in their development and your feedback on SAS Press books that you are using. Please visit sas.com/books to do the following:

- Sign up to review a book
- Recommend a topic
- Request information on how to become a SAS Press author
- Provide feedback on a book

Chapter 1: Understanding the Industry

In the pharmaceutical industry, there is a mandate to create standardized clinical data using very specific rules. These rules are created and governed by the Clinical Data Interchange Standards Consortium (CDISC). In this book, we describe and illustrate how to create these required CDISC data sets with SAS code. A statistical programmer should be familiar with the CDISC rules required to create standardized clinical trial data sets. After reading this book, readers will be able to understand CDISC standardized clinical data structures, as well as how to create it.

1.1 Statistical Programmer Work Process

In the pharmaceutical industry, the primary goal of a statistical programmer is to create standard data efficiently in order for the clinical trial biostatistician to perform their analysis. A simplified illustration of the process workflow for the statistical programmer is shown in Figure 1.1.

Figure 1.1 shows that the work process starts from the Case Report Form (CRF), which is designed for a specific study to collect clinical trial raw data from a site. Often, studies are global – having sites in countries all over the world. The Data Management group creates the CRF by working with the statistical programmer and other functions to ensure that the appropriate data is collected for the purpose of that study.

After the CRF is created and data is entered into it by the sites, the statistical programmer uses this data to create CDISC Study Data Tabulation Model (SDTM) domains to group collected information from the CRF in a way that facilitates standardization. The statistical programmer then creates CDISC Analysis Data Model (ADaM) data sets from the SDTM domains to support clinical trial analysis.

Figure 1.1: Statistical Programmer Process Workflow

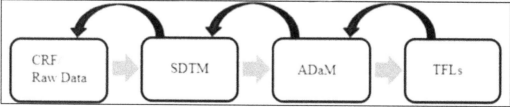

Note: When we refer to SDTM, we use the term domain, and for ADaM, we use the term data set. To be crystal clear, both models generate standardized clinical data using SAS.

Creating SDTM and ADaM data sets ensure that data will meet the criteria to be accepted by regulatory agencies such as the United States Food and Drug Administration (FDA). Finally, the statistical programmer generates the Tables, Figures, and Listings (TFLs), which are used to support analysis presented in the Clinical Study Report (CSR). The CSR is used to provide evidence to regulatory agencies about the safety and efficacy of the study drug.

Note: This workflow actually represents a much more complicated process. We intentionally keep it at a level where the reader can just focus on how the statistical programmer receives the raw data and creates the standardized clinical data (SDTM and ADaM).

1.2 Drug Approval Process

The FDA's Center for Drug Evaluation and Research (CDER or CBER for Biologics) reviews the SDTM and ADaM data created in the previous section to assess the drug's safety and efficacy. There are many stages of development and clinical trials as the drug approval process advances. The following are the most critical drug development terms and milestones:

- **Pre-clinical Studies**: Research often using animals to find out if a drug is likely to be safe in humans.
- **Investigational New Drug (IND) Application**: Facilitates permission to start clinical trials in humans if the pre-clinical study results are promising.
- **Phase I Clinical Trial**: First in human (FIH) study of a new drug, often looking at dose ranges, drug-drug interactions, food effect, etc.
- **Phase II Clinical Trial**: Explore and determine efficacy of a new drug.
- **End of Phase II Meeting**: Regulatory agency (for example, FDA) and sponsor agree on design of Phase III study.
- **Phase III Clinical Trial**: Large-scale clinical trial that confirms efficacy and safety that, if successful, will be reviewed by regulatory agencies for marketing approval.
- **Pre-NDA/BLA Review Meeting**: Discuss strategy for potential approval of the IND, format and content of the anticipated application, including labeling, risk evaluation and mitigation strategy, data structure and accessibility of data for submission.
- **New Drug/Biologic Application (NDA/BLA)**: New drug application that can lead to market approval, which allows the drug to be legally marketed.
- **Drug Labeling Review**: Identify drug contents, information, and specific warnings for administration, storage, and disposal.
- **Facility/Sponsor Inspection**: Regulatory agency visits the sponsor, sites, or manufacturing facilities to evaluate trial conduct and compliance with the protocol and other regulatory requirements. This is often performed after submission of an NDA/BLA.

- **Phase IV Clinical Trial**: Experiments to conduct the long-term safety of a new drug after the drug is approved and is on the market. These are often designed to meet approval or reimbursement in areas outside of the United States, Japan, and China. (All of these countries have their own regulatory agencies that require standardized data be submitted.)

Table 1.1: Summary Table for the Four Clinical Trial Phases

	Phase 1	**Phase 2**	**Phase 3**	**Phase 4 (pro-market)**
Participants	Healthy volunteers or patients	Patients	Patients	Patients
Number	20–100	Up to hundreds	300–3000	Large, diverse population
Length	From days up to several months	Several months–2 years	1–4 years	Several years
Goal	Safety and dosage	Efficacy and side effects	Efficacy and monitoring of adverse reactions (safety)	Long-term safety and efficacy
% Continuation	Around 70% of the drugs move to the next phase	Around 33% of drugs move to the next phase	Around 25–30% of drugs move to the next phase	

Source: FDA. https://www.fda.gov/patients/drug-development-process/step-3-clinical-research

1.3 Clinical Trial Study Design

The clinical trial design is one of the most critical interventional trial components. It serves to optimize the clinical trial conduct and provide the most objective range of approaches to evaluate the therapy. There are several clinical trial design concepts in practice that the reader needs to know. We only list the common clinical trial designs; in practice each one of these can be used in combination with each other. In addition, there are many other nuances to each design. For example, a Phase III randomized trial is often placebo-controlled and double-blind and has an open label extension for safety purposes.

- **Randomized:** Participants are divided randomly into separate treatment (placebo) groups that compare the groups.
- **Placebo-controlled:** Placebo is given to one group of participants, while a therapy is given to another group. Placebo is designed to have no real effect.
- **Open-label:** Both the researchers and the participants know which treatment is being administrated.

- **Double-blind:** Neither the participants nor the researchers know which treatment is assigned and administered.
- **Parallel Design:** Patients are randomly assigned to a treatment and remain on that treatment throughout the duration of the entire trial.
- **Crossover Design**: All subjects switch treatment regimens during the course of the trial.

For more information, please check the Drug Study Designs Guidance for Institutional Review Boards and Clinical Investigators: https://www.fda.gov/regulatory-information/search-fda-guidance-documents/drug-study-designs.

1.4 CDISC Standard Data Structures

CDISC is a global non-profit organization that develops data standards for the pharmaceutical industry. There are three distinct standard data models developed by CDISC for regulatory submissions that the reader needs to understand. The basic concepts for these three models are below. More details are provided in subsequent chapters.

- **Study Data Tabulation Model (SDTM):** Defines a standard structure for human clinical trial data tabulations that are sent to a regulatory authority such as the FDA as a part of the data submission package. The SDTM model is considered the 'raw' data for regulatory submission.
- **Analysis Data Model (ADaM):** Uses the SDTM domains to develop data sets for the purpose of summarizing and analyzing the clinical data. The ADaM model data sets generate all the analysis to support the trial.
- **Define-XML**: When sending SDTM and ADaM data sets to the regulatory authorities, it's critical to see the specifications and understand how to navigate the SDTM and ADaM data sets. DEFINE-XML provides a machine-readable version of how the SDTM and ADaM data sets were created, including any explanations about complex data derivations. This allows the FDA to work more efficiently with data submission.

1.5 Important Documents Summary

There are some key and important documents that are essential for statistical programmers to understand, use, or create in order to create standardized clinical data – SDTM and ADaM. We include TFLs as they are why we create SDTM and ADaM data sets. The documents are listed in the order they are created. There will be multiple iterations, and for the reader's sake, we don't feel it's necessary to talk about every single scenario. The only document that MUST be finalized before all other documents is the Protocol.

- **Protocol:** Detailed summary and guide of the study including study design, schedule of assessments, and analysis methods. Every subsequent document and the study conduct are based on the Protocol. It is reviewed and approved by Institutional Review Boards (IRBs), regulatory authorities, and sites.
- **Blank Case Report Form (CRF):** Used to collect all the information for every single patient in the study. The CRF is created by the data manager, then the statistical programmer, biostatisticians, and other functions review the CRF to confirm that all the data needed for analysis is captured. The CRF can only be finalized after the Protocol is finalized.
- **Statistical Analysis Plan (SAP):** Created by the study biostatistician and explains how the data is to be analyzed.
- **Table, Figure, and Listing templates (TFLs):** Created by the study biostatistician, these provide the content and detailed information to help statistical programmers create the actual TFLs once the SAP is stable.
- **SDTM Annotated Case Report Form (SDTM aCRF):** Annotated by the statistical programmer. Statistical programmers use the annotated CRF to generate and understand the structure of SDTM domains.
- **SDTM Specifications:** Provide details about how to generate the SDTM domains; they cover information about how to program all domains, including variables' lengths, labels, formats, as well as instructions on how to create each variable. They are created by the statistical programmer with the SDTM aCRF simultaneously as the two documents are highly correlated and dependent on each other.
- **ADaM Specifications:** Contain information about the analysis data sets from SDTM domains as well as new variables and derivations required for analysis purposes in ADaM data sets. These specifications are created by the statistical programmer. A stable SAP and TFL shells are required in order to generate ADaM specifications.
- **Define-XML:** Machine-readable version of specifications including the SDTM Define-XML document and ADaM Define-XML document. This also provides more detailed information about how the data was created.

Table 1.2: Summary Table of Important Documents

Document	Purpose	Statistical Programmers' Role	Time
Protocol	Detailed summary and guidance of the study	Study lead review	Before study starts
Blank CRF	Detailed data collection	Study lead review	Before study starts
SAP	Explain how the data is analyzed	Study lead review	Pre-Programming
TFL Templates	As a reference when creating tables, listings, and figures	Study lead review	Pre-Programming
SDTM aCRF	Link CRF with SDTM domains	Study lead creates	Pre-Programming

(Continued)

Table 1.2: (*Continued*)

Document	Purpose	Statistical Programmers' Role	Time
SDTM specifications	Explain the derivation of each variable in SDTM domains	Study lead creates	Pre-Programming
ADaM specifications	Explain the derivation of each variable in ADaM data sets	Study lead creates	Pre-Programming
Define-XML	Provides machine-readable version of specifications	Study lead creates	Study End

Chapter 2: Getting Started from the Case Report Form

An electronic Case Report Form (eCRF) is used to collect all clinical trial patient information. It's critical to understand that all of the values entered on the eCRFs need to be collected or converted to standard, pre-specified text, which is called controlled terminology (CT). In the Demographics domain, for example, the eCRF values for Sex are "Male" and "Female"; however, the CT for these are "M" and "F", respectively. The link to the CDISC Terminology is located here: https://datascience.cancer.gov/resources/cancer-vocabulary/cdisc-terminology.

The Data Management group drafts the CRF, which is then reviewed by the statistical programmer in conjunction with the entire cross-functional team, including members such as biostatisticians, medical monitors, and representatives from all relevant functions. The purpose of this cross-functional review is to confirm that every data point required for the study is captured accurately on the eCRF. Trained staff at the site enter patients' clinical trial information into the CRF.

In this chapter, we will show and discuss some of the most common and important standard eCRFs.

2.1 eCRF Portal

The link to the eCRFs from CDISC are located here: https://www.cdisc.org/kb/ecrf. Most of the data on each patient participating in a clinical trial is documented in the eCRF, including Demographics, Adverse Events, Concomitant Medications, Disposition data, and so on. The eCRF can vary by sponsor.

There are situations where the site might not have all available equipment to perform some examinations or sponsors have a preferred vendor to perform specific assessments. In these situations, this data will not be collected on the eCRF. For example, central lab data often comes from different sources and as such might be transferred separately in formats such as Excel.

2.2 Electronic CRFs (eCRFs)

As mentioned above, eCRFs are used in all clinical trials. In this section, we discuss key information for some essential standardized CDISC eCRFs, such as Demographics, Disposition, Adverse Events, Exposure, ECG, and Laboratory results.

Note: We reference CDISC eCRFs in this chapter for the sake of helping the reader understand the basic concepts of eCRFs. In other words, these eCRFs will contain the *minimum* amount of information collected on any eCRF in order to meet CDISC standards. In real life, the eCRF will be more customized.

2.2.1 Demographics

The Demographics (DM) eCRF collects the following information. (See Figure 2.1.)

- Birth Date: The value is collected in the following format: DD-MMM-YYYY.
  - Note: in some countries, due to patient privacy regulations, not all birthdate components can be collected.
- Age: This value is calculated as an integer using Informed Consent date and Birth Date. For example, if a patient is 18 years, 11 months, and 15 days old, the patient's age would be recorded as 18.
- Age Unit: This value is collected on the eCRF in years for adult studies and in months/ days for pediatric studies.
- Sex: Collected on the eCRF as "Male", "Female", "Unknown", and "Undifferentiated", and the corresponding CT are "M", "F", "U", and "UNDIFFERENTIATED".
- Ethnicity: Collected on the eCRF as "Hispanic or Latino", "Not Hispanic or Latino", "Not Reported", and "Unknown", and the corresponding CT are "HISPANIC OR LATINO", "NOT HISPANIC OR LATINO", "NOT REPORTED", and "UNKNOWN".
- Race: Collected on the eCRF as "White", "Black or African American", "Asian", "American Indian or Alaska Native", "Native Hawaiian or Other Pacific Islander", "Not Reported", "Unknown", or "Other". The CT are "W", "B", "A", "AA", "HP", "NOT REPORTED", "UNKNOWN", and "OTHER".
- Raceoth: Collected on the eCRF if the value of "Race" equals "Other". The value of "Other" will be what is recorded in "Specify Other Race".

Figure 2.1: Demographics (DM)

2.2.2 Disposition

The Disposition (DS) eCRF collects information about subject study and treatment completion, and if one or both are not completed, records the reason. (See Figure 2.2.)

- What was the subject's status at the end of the study: The values for this include "Adverse Event", "Completed", "Death", "Lost to Follow-Up", "Pregnancy", "Progressive Disease", "Protocol Deviation", "Screen Failure", "Site Terminated by Sponsor", "Study Terminated by Sponsor", "Withdrawal by Subject", and "Other". The corresponding CT are "ADVERSE EVENT", "COMPLETED", "DEATH", "LOST TO FOLLOW-UP", "PREGNANCY", "PROGRESSIVE DISEASE", "PROTOCOL DEVIATION", "SCREEN FAILURE", "SITE TERMINATED BY SPONSOR", "STUDY TERMINATED BY SPONSOR", "WITHDRAWAL BY SUBJECT", and "OTHER".
- Specify: The value for this is entered as free text if "Other" is selected in Question 1.1.
- What was the study discontinuation or completion date: The value should be in DD-MMM-YYYY format.

Figure 2.2: Disposition (DS)

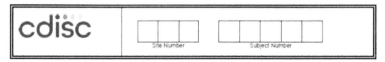

Form DS - Disposition			
1 DS - Disposition			
1.1	What was the subject's status at the end of study?	Adverse Event Completed Death Lost To Follow-Up Pregnancy Progressive Disease Protocol Deviation Screen Failure Site Terminated by Sponsor Study Terminated By Sponsor Withdrawal by Subject Other	DSDECOD
1.2	Specify		DSTERM
1.3	What was the study discontinuation or completion date? (DD-MMM-YYYY)		DSSTDAT

2.2.3 Adverse Events

The Adverse Events (AE) eCRF collects the following information. (See Figure 2.3.)

- Were any adverse events experienced: The values are "Yes" or "No".
  - Note: There are a few exceptions where variables collected on the eCRF are not created in SDTM domains as they not discussed in the SDTM IG, like *AEYN*. In these cases, the variables are annotated as "Not Submitted" on the eCRF.
- What is the adverse event term: The value is the text entered directly into the eCRF. It captures any unfavorable and unintended symptom during the study.
- Start Date: The value for adverse event start date in DD-MMM-YYYY format.
- Ongoing: The values are "Yes" or "No".
- End Date: The value for adverse event end date in DD-MMM-YYYY format.
- Severity: The values are "Mild", "Moderate", and "Severe". The corresponding CT are "1", "2", "3".
  Note: In many eCRFs, there are often five categories of severity reported. In addition to the three values listed above, there is also "Life-threatening" and "Death".
- Was the adverse event serious: The values are "Yes" and "No". If the adverse event was serious ("Yes"), then there are other subcategories including "Did the adverse event result in death", "Was the adverse event life threatening", "Did the adverse event result in disability or permanent damage", "Was the adverse event associated with a

congenital anomaly or birth defect" and "Was the adverse event a medically important event not covered by other serious problems".
- Relationship to Study Treatment: The values are "Yes" and "No". The corresponding CT are "Y" and "N".
- Action Taken with Study Treatment: The values can be a range of actions taken as a result of patient response to treatment, including "Dose Increased", "Dose Not Changed", "Dose Rate Reduced", "Dose Reduced", "Drug Interrupted", "Drug Withdrawn", "Not Applicable", and "Unknown".
  - Note: The corresponding CT are "DOSE INCREASED", "DOSE NOT CHANGED", "DOSE RATE REDUCED", "DOSE REDUCED", "DRUG INTERRUPTED", "DRUG WITHDRAWN", "NOT APPLICABLE", and "UNKNOWN".
- Other Action Taken: The values for this should include what is entered in this free text field.
- Outcome: The values here include "Fatal", "Not Recovered or Not Resolved", "Recovered or Resolved", "Recovered or Resolved with Sequelae", "Recovering or Resolving", and "Unknown". The corresponding CT are "NOT RECOVERED/NOT RESOLVED", "RECOVERED/RESOLVED", "RECOVERED/RECORVERED WITH SEQUELAE", "RECOVERING OR RESOLVING", and "UNKNOWN".

Figure 2.3: Adverse Events (AE)

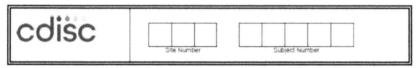

Form AE - Adverse Events			
1 AE - Adverse Events			
1.1	Were any adverse events experienced?	No Yes	AEYN
1.2	What is the adverse event term?		AETERM
1.3	Start Date (DD-MMM-YYYY)		AESTDAT
1.4	Ongoing	No Yes	AEONGO
1.5	End Date (DD-MMM-YYYY)		AEENDAT
1.6	Severity	Mild Moderate Severe	AESEV

| 1 7 | Was the adverse event serious? | ⊙ No   **AESER**
⊙ Yes

Did the adverse event result in death?   ⊙ No
          AESDTH   ⊙ Yes

Was the adverse event life threatening?   ⊙ No
          AESLIFE   ⊙ Yes

Did the adverse event result in initial or prolonged hospitalization for the subject?
     ⊙ No   **AESHOSP**
     ⊙ Yes

Did the adverse event result in disability or permanent damage?
     ⊙ No   **AESDISAB**
     ⊙ Yes

Was the adverse event associated with a congenital anomaly or birth defect?
     ⊙ No   **AESCONG**
     ⊙ Yes

Was the adverse event a medically important event not covered by other serious criteria?
     ⊙ No   **AESMIE**
     ⊙ Yes |
| 1 8 | Relationship to Study Treatment | ⊙ No   **AEREL**
⊙ Yes |

1 9	Action Taken with Study Treatment	⊙ Dose Increased **AEACN** ⊙ Dose Not Changed ⊙ Dose Rate Reduced ⊙ Dose Reduced ⊙ Drug Interrupted ⊙ Drug Withdrawn ⊙ Not Applicable ⊙ Unknown
1 10	Other Action Taken	**AEACNOTH**
1 11	Outcome	⊙ Fatal **AEOUT** ⊙ Not Recovered or Not Resolved ⊙ Recovered or Resolved ⊙ Recovered or Resolved with Sequelae ⊙ Recovering or Resolving ⊙ Unknown

2.2.4 Exposure

The Exposure (EX) eCRF captures the dose administration information for each subject shown below. (See Figure 2.4.)

- Study Treatment Label Identifier: The value is "EX Identifier".
- Start Date: The value is the start date of treatment in DD-MMM-YYYY format.

Figure 2.4: Exposure (EX)

Form EX - Exposure			
1 EX - Exposure			
1.1	Study Treatment Label Identifier	`EX.EXREFID`	***EXREFID***
1.2	Start Date (DD-MMM-YYYY)	`EX.EXSTDTC`	***EXSTDAT***
1.3	End Date (DD-MMM-YYYY)	`EX.EXENDTC`	***EXENDAT***
1.4	Dose	`EXDOSTXT or EXDOSE` ***EXDOSE***	***EXDOSTXT***
1.5	Units	Capsule Gram International Unit Microgram Milligram Milliliter Puff Tablet `EX.EXDOSU`	***EXDOSU***
1.6	Frequency	As Needed 4 Times per Day Three Times Daily Twice Daily Daily Every Other Day Every Month `EX.EXDOSFRQ`	***EXDOSFRQ***
1.7	Route	Intralesional Intramuscular Intraocular Intraperitoneal Nasal Oral Rectal Respiratory (Inhalation) Subcutaneous Topical Transdermal Vaginal `EX.EXDOROUTE`	***EXROUTE***

- End Date: The value is the end date of treatment in DD-MMM-YYYY format.
- Dose: The value is the dose amount.
- Units: The values include "Capsule", "Gram", "International Unit", "Microgram", "Milligram", "Milliliter", "Puff", and "Tablet". The corresponding CT are "CAPSULE", "GRAM", "INTERNATIONAL UNIT", "MICROGRAM", "MILLIGRAM", "MILLILITER", "PUFF", and "TABLET".
- Frequency: The values include "As Needed", "4 Times per Day", "Three Times Daily", "Twice Daily", "Daily", "Every Other Day", and "Every Month". The values of frequency vary widely, depending on the treatment regimen. The corresponding CT are "AS NEEDED", "4 TIMES PER DAY", "THREE TIMES DAILY", "TWICE DAILY", "DAILY", "EVERY OTHER DAY", and "EVERY MONTH".
- Route: The value is the route of administration for the intervention, which can be (not exhaustive): "Intralesional", "Intramuscular", "Intraocular", "Intraperitoneal", "Nasal", "Oral", "Rectal", "Respiratory", "Subcutaneous", "Topical", "Transdermal", and "Vaginal". The corresponding CT are "INTRALESIONAL", "INTRAMUSCULAR", "INTRAOCULAR", "INTRAPERITONEAL", "NASAL", "ORAL", "RECTAL", "RESPIRATORY", "SUBCUTANEOUS", "TOPICAL", "TRANSDERMAL", and "VAGINAL".

2.2.5 Concomitant Medications

The Concomitant Medications (CM) eCRF captures medications taken concurrently with investigational treatment, as shown below in Figure 2.5.

- Were any medications/therapies taken: The values can be "Yes" and "No".
  Note: There are a few exceptions where variables collected on the eCRF are not created in SDTM domains as they not discussed in the SDTM IG, like *CMYN*. In these cases, the variables are annotated as "Not Submitted" on the eCRF.
- Medication therapy: The value indicates what, if any, other medications/therapies were taken.
- Indication: The values can be for what indication the medication was taken for. The same medication can be taken for multiple indications, such as ibuprofen for headache as well as muscle ache, to name a few.
- Dose: The value for this is the dosage information. This is entered in free text and is intended to capture the amount of treatment taken.

Figure 2.5: Concomitant Medications (CM)

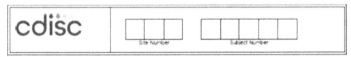

Site Number	Subject Number

Form CM - Concomitant Medications		
1 CM - Concomitant Medications Header		
1.1	Were any medications/ therapies taken?	○ No ○ Yes	CMYN
2 CM - Concomitant Medications		
2.1	Medication/Therapy		CMTRT
2.2	Indication		CMINDC
2.3	Dose		CMDSTXT
2.4	Unit	○ Percent Volume per Volume ○ Capsule ○ Gram ○ Inhalation ○ International Dosing Unit ○ Liter per Hour ○ Liter per Minute ○ Milligram ○ Milligram per Kilogram ○ Milliliter ○ Milliliter per Hour ○ Milliliter per Kilogram ○ Puff ○ Spray ○ Tablet ○ Microgram ○ Microgram per Kilogram ○ Other	CMDOSU
		Other, Specify [ ]	DOSUO

2	CM - Concomitant Medications		
2.5	Dose Form	⬚ Aerosol ⬚ Capsule ⬚ Cream ⬚ Gas ⬚ Gel ⬚ Inhalant ⬚ Injectable ⬚ Liquid ⬚ Ointment ⬚ Patch ⬚ Powder ⬚ Spray ⬚ Suppository ⬚ Suspension ⬚ Tablet ⬚ Other Other, Specify [_____]	CMDOSFRM DOSFRMO
2.6	Frequency	⬚ Twice Daily ⬚ Once ⬚ As Needed ⬚ Daily ⬚ Every Other Day ⬚ Three Times Daily ⬚ Unknown ⬚ Other Dose Frequency Other, Specify [_____] <center>DOSFRQO</center>	CMDOSFRQ
2.7	Route	⬚ Intralesional ⬚ Intramuscular ⬚ Intraocular ⬚ Intraperitoneal ⬚ Nasal ⬚ Oral ⬚ Rectal ⬚ Inhalation ⬚ Subcutaneous ⬚ Topical ⬚ Transdermal ⬚ Vaginal ⬚ Other Other, Specify [_____]	CMROUTE ROUTEO
2.8	Start Date (DD-MMM-YYYY)	[| | | | | | |]	CMSTDAT
2.9	Ongoing	⬚ No ⬚ Yes	CMONGO
2.10	End Date (DD-MMM-YYYY)	[| | | | | | |]	CMENDAT

- Unit: The value for this includes: "Percent Volume per Volume", "Capsule", "Gram", "Inhalation", "International Dose Unit", "Liter per Hour", "Liter per Minute", "Milligram", "Milligram per Kilogram", "Milliliter", "Milliliter per Hour", "Milliliter per Kilogram", "Puff", "Spray", "Tablet", "Microgram", "Microgram per Kilogram", and "Other". If the Unit is "Other", then specify what the other unit is. The corresponding CT are "PERCENT VOLUME PER VOLUME", "CAPSULE", "GRAM", "INHALATION", "INTERNATIONAL DOSE UNIT", "LITER PER HOUR", "LITER PER MINUTE", "MILLIGRAM",

"MILLIGRAM PER KILOGRAM", "MILLILITER", "MILLILITER PER HOUR", "MILLILITER PER KILOGRAM", "PUFF", "SPRAY", "TABLET", "MICROGRAM", "MICROGRAM PER KILOGRAM", and "OTHER".

- Dose Form: The value for this includes: "Aerosol", "Capsule", "Cream", "Gas", "Gel", "Inhalant", "Injectable", "Liquid", "Ointment", "Patch", "Powder", "Spray", "Suppository", "Suspension", "Tablet", and "Other". If the Dose form is "Other", then specify what Dose form is. The corresponding CT are "AEROSOL", "CAPSULE", "CREAM", "GAS", "GEL", "INHALANT", "INJECTABLE", "LIQUID", "OINTMENT", "PATCH", "POWDER", "SPRAY", "SUPPOSITORY", "SUSPENSION", "TABLET", and "OTHER".
- Frequency: The value for this includes: "Twice Daily", "Once", "As Needed", "Daily", "Every Other Day", "Three Times Daily", "Unknown", and "Other". If the Frequency is "Other", then specify what the Frequency is. This can vary based on a number of factors about how and why the medication is being used. The corresponding CT are "TWICE DAILY", "ONCE", "AS NEEDED", "DAILY", "EVERY OTHER DAY", "THREE TIMES DAILY", "UNKNOWN", and "OTHER".
- Route: The value of this is the route of administration for the Concomitant Medication and includes "Intralesional", "Intramuscular", "Intraocular", "Intraperitoneal", "Nasal", "Oral", "Rectal", "Inhalation", "Subcutaneous", "Topical", "Transdermal", "Vaginal", and "Other". If the Route is "Other", then specify what Route is. The corresponding CT are "INTRALESIONAL", "INTRAMUSCULSAR", "INTRAOCULAR", "INTRAPERITONEAL", "NASAL", "ORAL", "RECTAL", "INHALATION", "SUBCUTANEOUS", "TOPICAL", "TRANSDERMAL", "VAGINAL", and "OTHER".
- Start Date: The value is the start date of the concomitant medication in DD-MMM-YYYY format.
- Ongoing: The value for this is whether the treatment is ongoing or not. The values are "Yes" and "No", and the corresponding CT are "Y" and "N".
- End Date: The value for concomitant medication end date in DD-MMM-YYYY format.

2.2.6 Electrocardiogram

The Electrocardiogram eCRF captures various parameters derived from the EG waveform such as Heart Rate, PR Interval, QRS Duration, QT interval, and Interpretation for each subject as shown below in Figure 2.6. (Figure 2.6 includes central ECG and local ECG.)

Central ECG:

- Was an ECG performed: The values are "Yes" and "No".
  - Note: There are a few exceptions where variables collected on the eCRF are not created in SDTM domains as they not discussed in the SDTM IG, like *EGPERF*. In these cases, the variables are annotated as "Not Submitted" on the eCRF.
- ECG Reference Identifier/Accession Number: The value is "ECG Identifier".
- ECG Method: The value is the result of the 12 Lead Standard and Holter Continuous ECG recording.

- ECG Position: The values for this include "Sitting", "Standing", "Supine", "Semi-Recumbent", and "Semi-Fowlers". The corresponding CT are "SITTING", "STANDING", "SUPINE", "SEMI-RECUMBENT", and "SEMI-FOWLERS".
- ECG Date: The value is the date on which the ECG was performed in DD-MMM-YYYY format.

Local ECG:

- ECG Mean Heart Rate Result: The value is "beats/min". The corresponding CT is "beats/min".
- QRS Duration, Aggregate Result: The value is "msec". The corresponding CT is "msec".
- PR Interval, Single Beat Result: The value is "msec". The corresponding CT is "msec".
- QT Interval, Aggregate Result: The value is "msec". The corresponding CT is "msec".
- QTca Interval, Aggregate Result: The value is "msec". The corresponding CT is "msec".
- Interpretation: The values are either "Normal" or "Abnormal". If the interpretation is "Abnormal", then the subcategory is "Was the ECG clinically significant". The values are "Yes" or "No". The corresponding CR are "Y" and "N".

Figure 2.6: Electrocardiogram (EG)

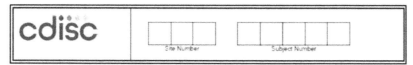

Form EG - Central Reading		
1 **EG - Central Reading**		
1.1	Was an ECG performed?	N No Y Yes	*EGPERF*
1.2	ECG Reference Identifier/ Accession Number		*EGREFID*
1.3	Method	12 LEAD STANDARD 12 LEAD STANDARD HOLTER CONTINUOUS ECG RECORDING HOLTER CONTINUOUS ECG RECORDING	*EGMETHOD*
1.4	Position	SITTING Sitting STANDING Standing SUPINE Supine SEMI-RECUMBENT Semi-Recumbent SEMI-FOWLERS Semi-Fowler's	*EGPOS*
1.5	Date (DD-MMM-YYYY)		*EGDAT*

Form EG - Local Reading		
1 EG - Local Reading		
1.1	Was an ECG performed?	No Yes *EGPERF*
1.2	Method	12 LEAD STANDARD *EGMETHOD* HOLTER CONTINUOUS ECG RECORDING
1.3	Position	Sitting *EGPOS* Standing Supine Semi-Recumbent Semi-Fowler's
1.4	ECG Date (DD-MMM-YYYY)	＼ ＼ ＼ ＼ ＼ ＼ ＼ *EGDAT*
1.5	ECG Mean Heart Rate Result	＼ ＼ ＼ beats/min *EGHRMN_EGORRES* *EGHRMN_EGORRESU*
1.6	QRS Duration, Aggregate Result	＼ ＼ ＼ msec *QRSAG_EGORRES* *QRSAG_EGORRESU*
1.7	PR Interval, Single Beat Result	＼ ＼ ＼ msec *PRSB_EGORRES* *PRSB_EGORRESU*
1.8	QT Interval, Aggregate Result	＼ ＼ ＼ msec *QTAG_EGORRES* *QTAG_EGORRESU*
1.9	QTca Interval, Aggregate Result	＼ ＼ ＼ msec *QTCAAG_EGORRES* *QTCAAG_EGORRESU*
1.10	Interpretation	Normal *INTP_EGORRES* Abnormal Was the ECG clinically significant? No *EGCLSIG* Yes

2.2.7 Lab

The Lab eCRF captures hematology, clinical chemistry, and urinalysis results such as Alkaline Phosphatase and Calcium for each subject. (See Figure 2.7.)

Central Lab:

- Was the sample collected: The values are "Yes" and "No".
  - Note: There are a few exceptions where variables collected on the eCRF are not created in SDTM domains as they not discussed in the SDTM IG, like *LBPERF*. In these cases, the variables are annotated as "Not Submitted" on the eCRF.
- Collection Date: The value is the central lab test collection date in DD-MMM-YYYY format.
- Collection Time: The value is the central lab test collection time in hh:mm (24-hour clock) format.

Local Lab:

- Laboratory Name: Name of the laboratory.
  - Note: There are a few exceptions where variables collected on the eCRF are not created in SDTM domains as they not discussed in the SDTM IG, like *LBNAM*. In these cases, the variables are annotated as "Not Submitted" on the eCRF.

- Was the sample collected: The values are "Yes" and "No".
  - Note: There are a few exceptions where variables collected on the eCRF are not created in SDTM domains as they not discussed in the SDTM IG, like *LBPERF*. In these cases, the variables are annotated as "Not Submitted" on the eCRF.
- Collection Date: The value is the local lab test collection date in DD-MMM-YYYY format.
- Collection Time: The format is hh:mm (24-hour clock).
- Was the subject fasting: The values are "Yes" and "No", and the corresponding CT are "Y" and "N".
- Alkaline Phosphatase (original result): The values are original Alkaline Phosphatase results.
- Alkaline Phosphatase (original unit): The values are "IU/L", "U/L", "ukat/L", and "umol/s/L".
- Normal Range Lower Limit of Alkaline Phosphatase: The value is the reference range lower limit in original units.
- Normal Range Higher Limit of Alkaline Phosphatase: The value is the upper limit in original units.
- Was the result clinically significant of Alkaline Phosphatase: The values are "Yes" and "No", and the corresponding CT are "Y" and "N".
- Calcium (original result): The values are original Calcium results.
- Calcium (original unit): The values are "mg/dL", "mEq/L", "mg/L", and "mmol/L".
- Normal Range Lower Limit of Calcium: The value is the reference range lower limit in original units.
- Normal Range Higher Limit of Calcium: The value is the reference range upper limit in original units.
- Was the result clinically significant of Calcium: The values are "Yes" and "No", and the corresponding CT are "Y" and "N".

Figure 2.7: Lab (LB)

Form LB - Central Processing		
1 LB - Central Processing		
1.1	Lab Panel Name	*LBCAT*
1.2	Was the sample collected?	☐ No ☐ Yes *LBPERF*
1.3	Collection Date (DD-MMM-YYYY)	*LBDAT*
1.4	Collection Time (24 hour clock)	*LBTIM*

Form LB - Local Processing				
1 LB - Local Processing				
1.1	Laboratory Name			**LBNAM**
1.2	Was the sample collected?	⬚ No ⬚ Yes		**LBPERF**
1.3	Collection Date (DD-MMM-YYYY)	⎢⎢⎢⎢⎢⎢⎢⎢⎢		**LBDAT**
1.4	Collection Time (24 hour clock)	⎢⎢⎢⎢		**LBTIM**
1.5	Was the subject fasting?	⬚ No ⬚ Yes		**LBFAST**

Alkaline Phosphatase **ALPU_LBORRES**	Alkaline Phosphatase Units **ALPU_LBORRESU**	Normal Range Lower Limit **LBORNRLO**	Normal Range Upper Limit **LBORNRHI**	Was this result clinically significant?
	⬚ IU/L ⬚ U/L ⬚ ukat/L ⬚ umol/s/L			⬚ No ⬚ Yes **LBCLSIG**

Calcium **CA_LBORRES**	Calcium Units **CA_LBORRESU**	Normal Range Lower Limit **LBORNRLO**	Normal Range Upper Limit **LBORNRHI**	Was this result clinically significant?
	⬚ mg/dL ⬚ mEq/L ⬚ mg/L ⬚ mmol/L			⬚ No ⬚ Yes **LBCLSIG**

2.3 Annotating the eCRF

An annotated Case Report Form (aCRF) is an eCRF in PDF format that provides specific information about SDTM domain names, variable names as well as the values that variables can be populated with in order to meet CDISC SDTM standards. In other words, the annotation provided on the aCRF is for SDTM creation purposes. The reader might be wondering why the CDISC-required information isn't collected in CDISC format on the eCRF. The simple answer is that it is collected in CDISC format as much as possible; however, the link between eCRF standards and SDTM standards is not widely implemented across the industry at this time.

FDA guidance recommends submitting the aCRF when the protocol is submitted. The aCRF is an important document to provide a traceable and transparent description of the data.

Note: There is only annotation for SDTM as ADaM is based solely on SDTM and any external data not collected on the CRF.

The aCRF annotation should align with the current CDISC SDTM-Metadata Submission Guideline (SDTM-MSG) document from the FDA, as described below. The CDISC SDTM_MSG document provides the necessary information recommended for electronic submissions. The following sections contain the central takeaways of aCRF annotation.

2.3.1 Annotating Unique eCRF Pages

It is recommended that sponsors include and annotate unique eCRF forms only. For example, a patient will only complete the Demographics page once, while they will most likely have multiple Vital Signs, Concomitant Medications, and Adverse Event forms. Vital Signs are scheduled assessments while CM and AE are entered any time a patient starts taking a new medication or has a new AE. By using the bookmarking tool in PDF, this will represent the form as many times as was intended in the Schedule of Assessments (SOA).

2.3.2 Appearance of Annotations

Each domain represented on the CRF page or collection screen should have its own annotation. The following are guidelines for the appearance of annotations: domain annotations should use black text with bold formatting; variable annotations should use black text without bold formatting; annotations should use a 12-point font size, except where a sponsor chooses to increase or reduce the [annotation] font size to accommodate their CRF pages or collection screens required in order to comply with regulatory authority guidance/requirements. Lastly, annotations for variables and data set codes should be capitalized, and instructional text and comments should be sentence case, excluding variables and domain codes, which should be capitalized as indicated above.

The CDISC MSG recommends using the following sequence of colors (blue, yellow, green, orange) when annotating multiple domains on a single eCRF page, as shown below. Figure 2.8 shows the sequence of colors for aCRF. Notes, which are annotations explaining a situation on the CRF and not direct variable annotations, should have the color of the domain to which they belong.

We have summarized the key takeaways from the CDISC SDTM-MSG. However, we *insist* that readers review the most current version of CDISC and FDA guidance documents, which will help statistical programmers annotate every SDTM domain consistently using the best, current guidance.

Figure 2.8: Sequence of Colors for aCRF

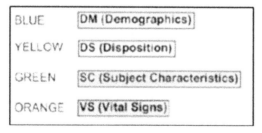

https://www.cdisc.org/standards/foundational/sdtmig/metadata-submission-guideline-msg-sdtmig (Study Data Tabulation Model Metadata Submission Guidelines) https://www.fda.gov/media/131872/download (Study Data Technical Conformance Guide)

https://www.fda.gov/media/76797/download (Portable Document Format (PDF) Specifications)

2.4 Annotated CRF Practices

In this section, we present the annotated CRFs for the corresponding eCRFs, which were listed in Section 2.1.

Note: Some variables in the CRF are not recommended to be submitted, so we marked "NOT SUBMITTED" in CRF pages.

Figure 2.9: Annotated Demographics (DM)

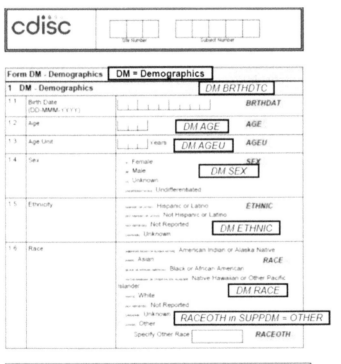

Figure 2.10: Annotated Disposition (DS)

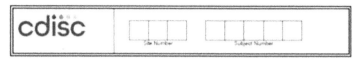

Form DS - Disposition

1 DS - Disposition | DS = Disposition

1.1	What was the subject's status at the end of study?	Adverse Event — Completed — Death — Lost To Follow-Up — Pregnancy — Progressive Disease — Protocol Deviation — Screen Failure — Site Terminated by Sponsor — Study Terminated By Sponsor — Withdrawal by Subject — Other	**DSDECOD** DS.DSDECOD
1.2	Specify	DSDECOD=DSTERM	**DSTERM**
1.3	What was the study discontinuation or completion date? (DD-MMM-YYYY)		**DSSTDAT** DS.DSSTDTC

Figure 2.11: Annotated Adverse Events (AE)

Form AE - Adverse Events

1 AE - Adverse Events

1.1	Were any adverse events experienced?	No — AEYN / Yes — NOT SUBMITTED	**AEYN**
1.2	What is the adverse event term?	AE.AETERM	**AETERM**
1.3	Start Date (DD-MMM-YYYY)		**AESTDAT** AE.AESTDTC
1.4	Ongoing	No — AE.AEONGO / Yes — NOT SUBMITTED	**AEONGO**
1.5	End Date (DD-MMM-YYYY)		**AEENDAT** AE.AEENDTC
1.6	Severity	Mild — Moderate — Severe	**AESEV** AE.AETOXGR

Figure 2.12: Annotated Exposure (EX)

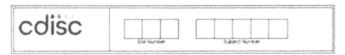

Form EX - Exposure			
1 EX - Exposure			
1.1	Study Treatment Label Identifier	EX.EXREFID	EXREFID
1.2	Start Date (DD-MMM-YYYY)	EX.EXSTDTC	EXSTDAT
1.3	End Date (DD-MMM-YYYY)	EX.EXENDTC	EXENDAT
1.4	Dose	EXDOSTXT or EXDOSE EXDOSE	EXDOSTXT
1.5	Units	Capsule / Gram / International Unit / Microgram / Milligram / Milliliter / Puff / Tablet EX.EXDOSU	EXDOSU
1.6	Frequency	As Needed / 4 Times per Day / Three Times Daily / Twice Daily / Daily / Every Other Day / Every Month EX.EXDOSFRQ	EXDOSFRQ
1.7	Route	Intralesional / Intramuscular / Intraocular / Intraperitoneal / Nasal / Oral / Rectal / Respiratory (inhalation) / Subcutaneous / Topical / Transdermal / Vaginal EX.EXDOROUTE	EXROUTE

Figure 2.13: Annotated Concomitant Medications (CM)

Form CM - Concomitant Medications			
1	**CM - Concomitant Medications Header**		
1 1	Were any medications/ therapies taken?	☐ No ☐ Yes	*CMYN* **CMYN** NOT SUBMITTED
2	**CM - Concomitant Medications**		
2 1	Medication/Therapy		*CM CMTRT* **CMTRT**
2 2	Indication		*CM CMINDC* **CMINDC**
2 3	Dose		*CM CMDSTXT* **CMDSTXT**
2 4	Unit	☐ Percent Volume per Volume ☐ Capsule ☐ Gram ☐ Inhalation ☐ International Dosing Unit ☐ Liter per Hour ☐ Liter per Minute ☐ Milligram ☐ Milligram per Kilogram ☐ Milliliter ☐ Milliliter per Hour ☐ Milliliter per Kilogram ☐ Puff ☐ Spray ☐ Tablet ☐ Microgram ☐ Microgram per Kilogram ☐ Other Other, Specify ☐	**CMDOSU** *CM CMDOSU* *DOSUO in SUPPCM* **DOSUO**

Figure 2.14: Annotated Electrocardiogram (EG)

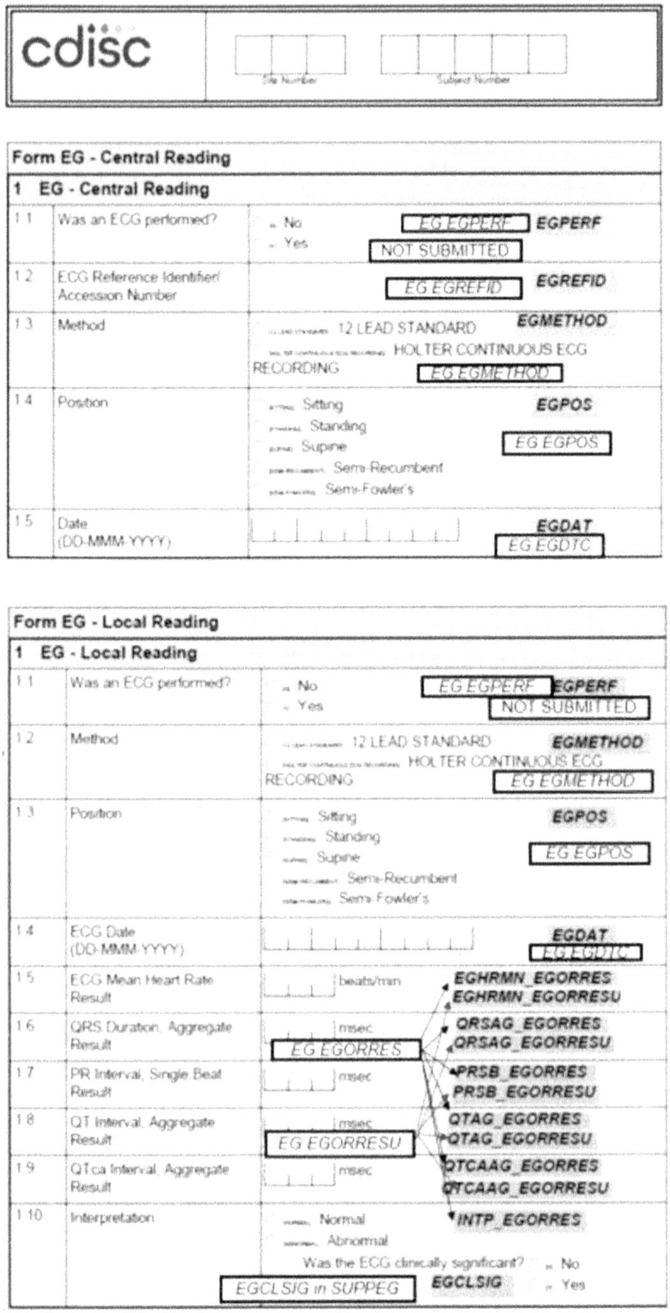

Figure 2.15: Annotated Laboratory Test Data (LB)

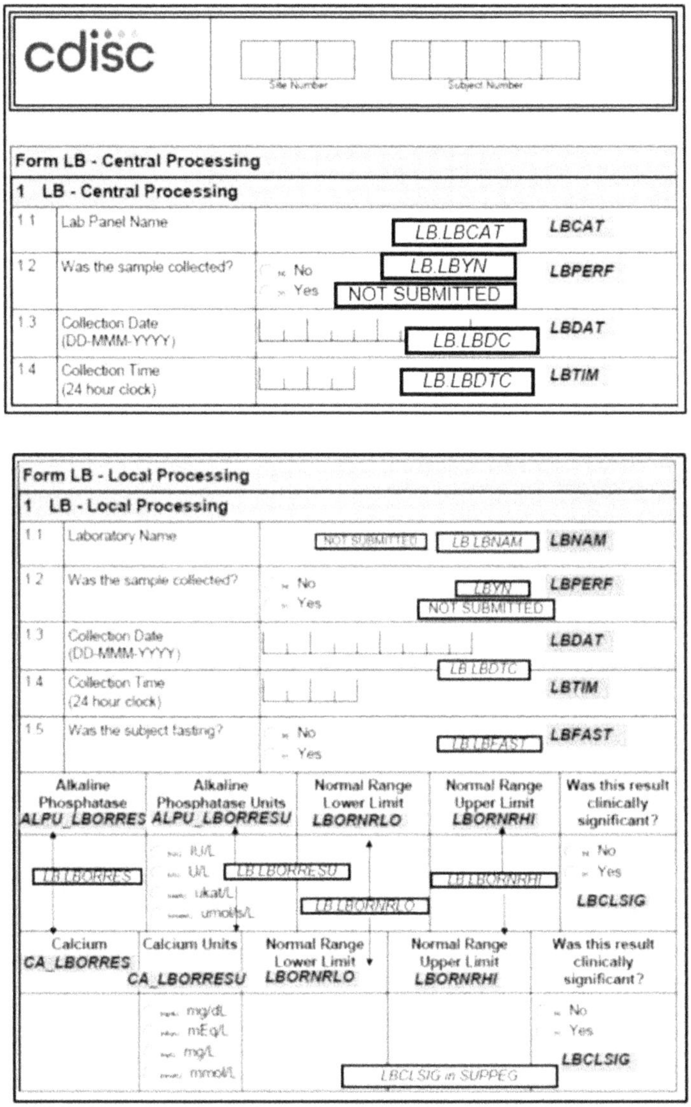
cdisc

Site Number | Subject Number

Form LB - Central Processing

1 LB - Central Processing

1.1	Lab Panel Name	LB.LBCAT	LBCAT
1.2	Was the sample collected?	No / Yes — LB.LBYN — NOT SUBMITTED	LBPERF
1.3	Collection Date (DD-MMM-YYYY)	LB.LBDC	LBDAT
1.4	Collection Time (24 hour clock)	LB.LBDTC	LBTIM

Form LB - Local Processing

1 LB - Local Processing

1.1	Laboratory Name	NOT SUBMITTED — LB.LBNAM	LBNAM
1.2	Was the sample collected?	No / Yes — LBYN — NOT SUBMITTED	LBPERF
1.3	Collection Date (DD-MMM-YYYY)	LB.LBDTC	LBDAT
1.4	Collection Time (24 hour clock)		LBTIM
1.5	Was the subject fasting?	No / Yes — LB.LBFAST	LBFAST

Alkaline Phosphatase ALPU_LBORRES	Alkaline Phosphatase Units ALPU_LBORRESU	Normal Range Lower Limit LBORNRLO	Normal Range Upper Limit LBORNRHI	Was this result clinically significant?
LB.LBORRES	IU/L, U/L, ukat/L, umol/s/L — LB.LBORRESU	LB.LBORNRLO	LB.LBORNRHI	No / Yes LBCLSIG

Calcium CA_LBORRES	Calcium Units CA_LBORRESU	Normal Range Lower Limit LBORNRLO	Normal Range Upper Limit LBORNRHI	Was this result clinically significant?
	mg/dL, mEq/L, mg/L, mmol/L			No / Yes LBCLSIG

LBCLSIG in SUPPEG

Chapter 3: Study Data Tabulation Model (SDTM)

The Study Data Tabulation Model (SDTM) is a standard structure for human clinical trial data. The data is referred to as a domain and is required as part of a drug or biological submission to a regulatory authority such as the United States Food and Drug Administration (FDA). All SDTM domains are built around the concept of creating observations collected about subjects who participated in a clinical study. The domains are unique, created around topics and comprised of two letters such as DM for Demographics domain, AE for Adverse Events domain, etc., and reflect what has been entered into the CRF. In this chapter we will describe the SDTM data model, including information about the variables, standard domains as well as "role" and "core" variables, which exist in every SDTM domain. We will then discuss a very important and fundamental concept of clinical trials – the clinical trial schedule of assessments. Lastly, we will discuss a specific class of SDTM domains – called "Trial Design".

3.1 Variable "Roles"

Each observation can be described by a group of variables, corresponding to a row in a domain. A role determines the type of information conveyed by the variable about each specific observation and how it can be used. Variables can be classified into five major roles, referenced in CDISC's *Study Data Tabulation Model Implementation Guide (SDTMIG):*

1. Identifier variables: Study, unique subject, domain, and sequence number identifiers.
2. Topic variables: Focus of the observation.
3. Timing variables: Timing description, such as start date and end date of an event.
4. Qualifier variables: Values that describe the results or additional traits of the observation (such as units).
5. Rule variables (Trial Design domains): Values that describe the study and study conduct.

Note: In each SDTM domain, it's required to order the variables as follows: identifier, topic, qualifier, and timing variables.

3.2 SDTM Standard Domains

The SDTM standard domains can be classified as two groups: special purpose and general observation. Special purpose domains have their own distinct structure and cannot include additional variables other than what is specified in the SDTM IG. The general observation domains (other than special purpose domains) are divided into three classes: Interventions, Events, and Findings.

Special Purpose Domains:

- Demographics (DM); Comments (CO); Subject Elements (SE); Subject Visits (SV).
- Trial Design:
  - Trial Arm (TA); Trial Elements (TE); Trial Inclusion/Exclusion (TI); Trial Summary (TS); Trial Visits (TV).

General Observation Domains:

- Interventions:
  - Concomitant Medications (CM); Exposure as Collected (EC); Exposure (EX); Procedures (PR); Substance Use (SU).

- Events:
  - Adverse Events (AE); Clinical Events (CE); Disposition (DS); Protocol Deviations (DV); Healthcare Encounters (HO); Medical History (MH).

- Findings (As the findings domains are numerous and structured in the same manner, we've included the most common.):
  - ECG Test Results (EG); Laboratory Test Results (LB).

3.3 SDTM Core Variables

The concept of core variables is used both as a measure of compliance and to provide general guidance to sponsors. There are three categories of core variables: required, expected, and permissible.

- Required variables: Essential to the identification of a data record to make it meaningful.
  - Must be included in the domain and cannot have null values.

- Expected variables: Necessary to make a record useful.
  - Must also be included in the domain; however, there can be null values.

- Permissible variables: Optional in the domain as appropriate. Timing variables and qualifier variables from the general observation domains are considered permissible variables.
  - Note: For every SDTM domain, there could be information captured on the eCRF that does not fall into the three categories mentioned above. In those cases, we will create a SUPP-*domain name* domain, which will be described in Section 3.5, "Creating a New Domain."

3.4 Clinical Trial Schedule of Assessments

There are always three assessment stages in the clinical trial: 1st: "Screening," 2nd: "Treatment," 3rd: "Follow-Up."

- Screening: Period in which potential subjects are screened to determine whether they are eligible to enroll in the study.
- Treatment: Period in which researchers find out if the new drug is safe and effective for subjects.
- Follow-up: Period in which subjects finish the treatment period and are monitored until treatment effect wears off.

3.5 Creating a New Domain

In order to create a new CDISC-compliant domain, we need to select and populate the Required Identifier Variables, such as *STUDYID, DOMAIN*; Timing Variables, such as Start and End dates (*XX–STDTC, XX-ENDTC*, respectively); Topic Variables such as *SUBJID*; and Qualifier Variables based on the general observation class, such as *XX–TESTCD*.

Figure 3.1: The Process of Creating a New Domain

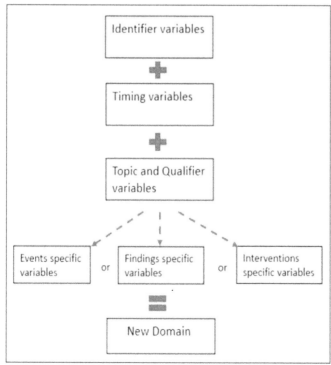

In our day-to-day work, we don't follow an exact linear process as shown in Figure 3.1 for creating identifier variables, then timing variables, and lastly topic and qualifier variables as it might not be possible to create all timing variables before topic and qualifier variables. In the following pages, we will demonstrate why this is, as well as how to create domains.

3.6 Model for SDTM Generation

3.6.1 Demographics (DM)

The DM domain is very important as it is used in all other SDTM domains and includes details about a set of essential, standard variables such as unique subject identifier, treatment and/or study start and end dates, the planned and actual treatment regimens, and other key variables that describe each subject's experience as it relates to the clinical study.

Structure of DM

The structure of DM consists of Identifier, Timing, Topic, and Qualifier variables, as follows.

Identifier Variables:

- STUDYID: Required variable in character format the unique study identifier. Each study has a unique study ID. In the example study described in this book, we populate it with "ABC-001".
- DOMAIN: Required variable in character format. It must be two characters such as "AE", "CM", "DM", etc.
- USUBJID: Required variable in character format used to uniquely identify a subject across all studies. The USUBJID variable is derived using: STUDYID- SITED-SUBJID, such as "ABC-001-001-001".

Topic Variables:

- SUBJID: Required variable in character format. The subject identifier must be unique within the study.

Qualifier Variables:

- SITEID: Required variable in character format. The study site identifier is always used to populate *USUBJID*.
- ARM: Required variable in character format. The description of the planned arm (the treatment subject) was assigned.
  - Note: In some cases, it's possible the subject is randomized into a treatment arm but, for any number of reasons, receives a different treatment arm. In those cases, it's critical to note that the planned and actual treatment arm differ.

- ARMCD: Required variable in character format and the code for the arm (above).
- ACTARMCD: Required variable in character format, the code of actual arm.
- ACTARM: Description of actual arm.
  - Note: For the purposes of this book, ACTARM equals ARM. We would like to point out that *ARM/ARMCD* are the same as *ACTARM* and *ACTARMCD* most of the time. However, there will be instances where a subject is randomized to a treatment arm but takes another treatment, resulting in the *ACTARM* and *ACTARMCD*, respectively.

- COUNTRY: Required in character format, the country where the subject participated in the trial.

  Note: For treatment planned/actual arm, we use the convention of capitalizing ARM when it is referred to in the context of a variable.

DM eCRF Reference

Demographic information is collected in the "Screening" stage. In practice, some demographic information is recorded in multiple CRF pages. In our simulated study, we use the standard simplified demographics eCRF from CDISC. We recommend readers look at the eCRF page in Chapter 2 before moving forward.

Figure 3.2: SDTM Annotated Demographics (DM)

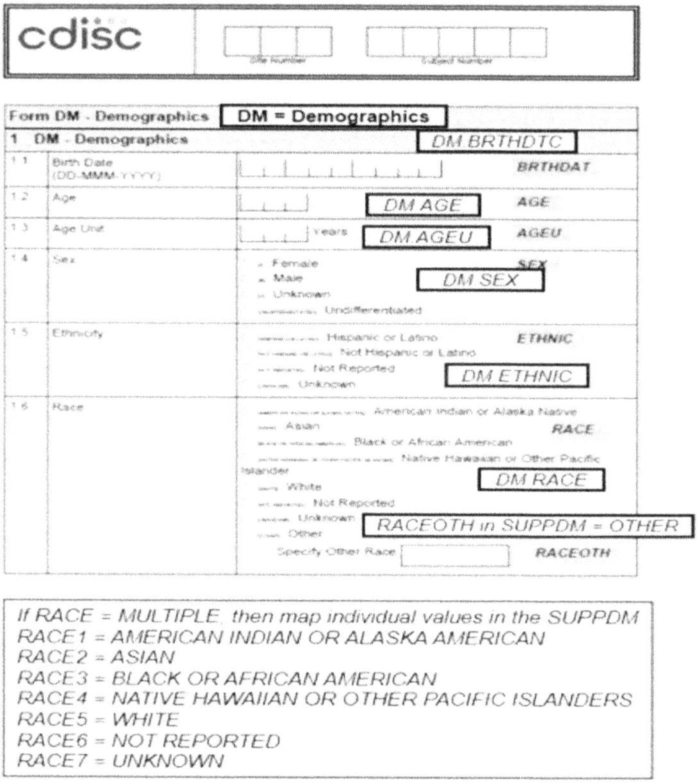

DM-SUPPDM Specification

Table 3.1: DM Specification

Variable	Label	Type	Length	Control Terminology	Origin	Notes
Unique identifier for study (STUDYID)	Study Identifier	char	20		Protocol: Set to "ABC-001"	This information that programmer can find in multiple raw data sets, such as RAW.IC; RAW.DM; RAW.AE, etc.
DOMAIN	Domain Abbreviation	char	2	DOMAIN	Assigned: Set to "DM"	
USUBJID	Unique Subject Identifier	char	40		Derived: Concatenate among three variables below: STUDYID-SITED-SUBJID	
Subject identifier that must be unique within the study (SUBJID)	Subject Identifier for the study	char	20		CRF	This information that programmer can find in multiple raw data sets, such as RAW.IX, RAW.DM, RAW.AE, etc.
RFSTDTC	Subject Reference Start Date/Time	char	20		Derived: Date and time of first dosing for every subject	The minimum value of EX.EXSTDAT
RFENDTC	Subject Reference End Date/Time	char	20		Derived: Date and time of last dosing for every subject	The maximum value of EX.EXSTDAT

(Continued)

Table 3.1: (*Continued*)

Variable	Label	Type	Length	Control Terminology	Origin	Notes
SITED	Study Site Identifier	char	10		eDT Collected in EDC but not on CRF	Raw DM data set, variable SITE
BRTHDTC	Date/Time of Birth	char	20		CRF	Raw DM data set, variable BRTHDAT
AGE	Age	num	8		CRF	Raw DM data set, variable AGE
AGEU	Age Units	char	10		CRF	Raw DM data set, variable AGEU
SEX	Sex	char	2		CRF	Raw DM data set, variable SEX
RACE	Race	char	100		CRF:	If subjects have multiple races, then DM.RACE="MULTIPLE";
ETHNIC	Ethnicity	char	60		CRF	Raw DM data set, variable ETHNTC
ARM	Description of Planned ARM	char	200		Derived: TA.ARM when TA.ARMCD= DM.ARMCD.	TA.ARMCD
ARMCD	Planned ARM Code	Char	20	ARMCD* ACTARMCD	Assigned: For the open label study, Set to "DRUG A"	
ACTARMCD	Actual ARM Code	char	20		Derived: ACTARMCD= ARMCD	
ACTARM	Description of Actual ARM	char	200		Equal to TA.ARM	TA.ARM

(*Continued*)

Table 3.1: (*Continued* **)**

Variable	Label	Type	Length	Control Terminology	Origin	Notes
COUNTRY	Country	char	4	COUNTRY	eDT: Collected in EDC but not in CRF	From raw DM data set, COUNTRY variable, then map with Control Terminology value

Table 3.2: SUPPDM Specification

Variable	Label	Type	Length	Control Terminology	Origin	Notes
STUDYID	Study Identifier	char	20		Protocol: Set to "ABC-001"	
RDOMAIN	Related Domain Abbreviation	char	2	DOMAIN	Assigned: Set to "SUPPDM"	
USUBJID	Unique Subject Identifier	char	40		Derived: Concatenate among three variables below: STUDYID-SITED-SUBJID	
IDVAR	Identifying variable	char	8		Assigned	
IDVARVAL	Identifying variable value	char	40		Assigned	
QNAM	Qualifier variable name	char	40		Assigned	
QLABEL	Data Value	char	40		Assigned	
QVAL	Indication	char	100			

(*Continued*)

Table 3.2: (*Continued*)

Variable	Label	Type	Length	Control Terminology	Origin	Notes
QNAM EQ "RACE1"	American Indian/ Alaska Native			Race	CRF	Set to "AMERICAN INDIAN OR ALASKA NATIVE" when DM.RACE= "MULTIPLE" and DM. RACE_ AINDIAN is not missing.
QNAM EQ "RACE2"	**ASIAN**			Race	CRF	Set to "ASIAN" when DM.RACE= "MULTIPLE" and DM.RACE_ASIAN is not missing.
QNAM EQ "RACE3"	Black or African American			Race	CRF	Set to "BLACK OR AFRICAN AMERICAN" when DM.RACE= "MULTIPLE" and DM.RACE_BLACK is not missing.
QNAM EQ "RACE4"	Native Hawaiian/ Pacific Islander			Race	CRF	Set to "NATIVE HAWAIIAN OR OTHER PACIFIC ISLANDER" when DM.RACE= "MULTIPLE" and DM.RACE_ HAWAIIAN is not missing.
QNAM EQ "RACE5"	White			Race	CRF	Set to "WHITE" when DM.RACE= "MULTIPLE" and DM.RACE_WHITE is not missing.
QNAM EQ "RACE6"	Not reported			Race	CRF	Set to " NOT REPORTED" when DM.RACE= "MULTIPLE" and DM.RACE_ NOREPORT is not missing.

(*Continued*)

Table 3.2: (*Continued*)

Variable	Label	Type	Length	Control Terminology	Origin	Notes
QNAM EQ "RACE7"	Unknown			Race	CRF	Set to "UNKNOWN" when DM.RACE= "MULTIPLE" and RACE_UNKNOWN is not missing.
QNAM EQ "RACE8"	Other			Race	CRF	Set to " " when DM.RACE= "MULTIPLE" and RACE_OTHER is not missing.
QORIG	Origin	char	100		Assigned	Set to "CRF"

Note: In order to create the SDTM.DM, we need the RAW.DM data set and other raw data sets including EX to derive RFSTDTC, and RAW.TA data set to derive *ARM* and *ARMCD*.

DM Programming

In order to create SDTM.DM (and all other SDTM domains), we first run PROC CONTENTS on the raw data. Please see code below.

Note: For every domain, the corresponding SAS program name is domain.sas. For example, we name the program dm.sas to create SDTM.DM, so all code in this section belongs to dm.sas.

```
/*Begin writing SAS program dm.sas*/
/*show structure of the raw DM dataset*/
proc contents data=RAW.DM;
run;
```

First, as seen in Figure 3.3, we use PROC CONTENTS to see that there are 18 variables in the raw DM data set. Further, we can see the Type, Length (Len), Format, and Informat of each variable. These variable attributes are assigned by SAS based on the raw data format. The Label is assigned in SAS based on the first row of the raw DM data set in the Excel sheet.

Note: The purpose of running PROC CONTENTS is to get a clear idea of what the raw metadata looks like and compare how similar it is to the SDTM domain that we will create from it.

Figure 3.3: Alphabetic List of Variables and Attributes

Alphabetic List of Variables and Attributes						
#	Variable	Type	Len	Format	Informat	Label
7	AGE	Num	8	BEST		AGE
8	AGEU	Char	5	$5	$5	AGEU
6	BRTHDAT	Num	8	MMDDYY10		BRTHDAT
2	COUNTRY	Char	13	$13	$13	COUNTRY
10	ETHNIC	Char	22	$22	$22	ETHNIC
5	FORM	Char	12	$12	$12	FORM
15	RACE_AINDIAN	Char	32	$32	$32	RACE_AINDIAN
14	RACE_ASIAN	Char	1	$1	$1	RACE_ASIAN
13	RACE_BLACK	Char	25	$25	$25	RACE_BLACK
12	RACE_HAWAIIAN	Char	1	$1	$1	RACE_HAWAIIAN
16	RACE_NOREPORT	Char	1	$1	$1	RACE_NOREPORT
18	RACE_OTHER	Char	1	$1	$1	RACE_OTHER
17	RACE_UNKNOWN	Char	1	$1	$1	RACE_UNKNOWN
11	RACE_WHITE	Char	5	$5	$5	RACE_WHITE
9	SEX	Char	6	$6	$6	SEX
3	SITE	Char	3	$3	$3	SITE
1	STUDYID	Char	7	$7	$7	STUDYID
4	SUBJID	Char	3	$3	$3	SUBJID

We use the information from the PROC CONTENTS as well as the SDTM IG and any sponsor-specific requirements in order to generate the SDTM data.

Note: For the purpose of enabling the reader to create the SAS data sets, we have provided spreadsheets and SAS code to import the spreadsheets in the Appendix. Other than the Appendix, we won't cover Excel and will instead focus on the raw SAS data, which is more realistic for working in the pharmaceutical industry.

Note: For ease of reading and creation of the code, we have separated each program into sections (using the /* and */ characters) to clearly document what and how variables are being created. For more information about why we discuss these variables, please refer to Section 3.3.

```
/*Create the 1st set of DM variables using existing variables from RAW.DM*/
data DM1;
  /*Specify length for standard variables*/
  length STUDYID ARMCD $20 ETHNIC $60 SEX $2 COUNTRY $4 BRTHDTC $20 RACE
$100; ❶
  set RAW.DM (rename=(COUNTRY=COUNTRY_ SEX=SEX_ AGEU=AGEU_ ETHNIC=ETHNIC_));

  /*Derive SITEID, BRTHDTC and COUNTRY*/
```

```
   SITEID=SITE;
   BRTHDTC=put(BRTHDAT,yymmdd10.);              ❷
   if COUNTRY_="United States" then COUNTRY="USA";

   /*Derive SEX*/
   if SEX_="Female" then SEX="F";
   else if SEX_="Male" then SEX="M";
   else if SEX_="Unknown" then SEX="U";
   else if SEX_="Undifferentiated" then SEX="UNDIFFERENTIATED";

   /*Derive ETHNIC AND AGEU*/

   ETHNIC=upcase(ETHNIC_);
   AGEU=upcase(AGEU_);
   /*Derive RACE*/

   if cmiss(RACE_WHITE, RACE_BLACK, RACE_HAWAIIAN, RACE_ASIAN,     ❸
RACE_AINDIAN,RACE_NOREPORT, RACE_UNKNOWN, RACE_OTHER)=7 then do;
     if not missing(RACE_AINDIAN) then RACE="AMERICAN INDIAN OR ALASKA
AMERICAN";
     else if not missing(RACE_ASIAN) then RACE="ASIAN";
     else if not missing(RACE_BLACK) then RACE="BLACK OR AFRICAN AMERICAN";
     else if not missing(RACE_HAWAIIAN) then RACE="NATIVE HAWAIIAN OR OTHER
PACIFIC ISLANDERS";
     else if not missing(RACE_WHITE) then RACE="WHITE";
     else if not missing(RACE_NOREPORT) then RACE="NOT REPORTED";
     else if not missing(RACE_UNKNOWN) then RACE="UNKNOWN";
     else if not missing(RACE_OTHER) then RACE=" ";
   end;

   else if 8-cmiss(RACE_WHITE, RACE_BLACK, RACE_HAWAIIAN, RACE_ASIAN, RACE_
AINDIAN,RACE_NOREPORT, RACE_UNKNOWN, RACE_OTHER)>1 then RACE="MULTIPLE";
   ❹

   /*Create SUPPDM Domain*/
   if RACE="MULTIPLE" then do;              ❺
     if not missing(RACE_AINDIAN) then RACE1="AMERICAN INDIAN OR ALASKA
AMERICAN";
     else if not missing(RACE_ASIAN) then RACE2="ASIAN";
     else if not missing(RACE_BLACK) then RACE3="BLACK OR AFRICAN AMERICAN";
     else if not missing(RACE_HAWAIIAN) then RACE4="NATIVE HAWAIIAN OR OTHER
PACIFIC ISLANDERS";
     else if not missing(RACE_WHITE) then RACE5="WHITE";
     else if not missing(RACE_NOREPORT) then RACE6="NOT REPORTED";
     else if not missing(RACE_UNKNOWN) then RACE7="UNKNOWN";
     else if not missing(RACE_OTHER) then RACE8=" ";
   end;

   if not missing(race_other) then RACEOTH="OTHER";

   ARMCD="DRUG A";
run;
```

❶ Define the variables' lengths to the corresponding SDTM IG.

❷ Be diligent when you create variables with date and time, such as BIRTHDTC, which is derived from the variable BRTHDAT (numeric) in the RAW.DM data set. We use put() with format *yymmdd10.* to convert numeric value to character in *yyyy-mm-dd.* format.

❸ We use cmiss() to count character and numeric missing values. First, we use cmiss () to test if there are seven missing variables out of the eight in total, and specifically which one is the not missing variable. RACE is equal to that nonmissing variable. If more than two variables are not missing (8-cmiss(XX)>1), then *RACE* is "MULTIPLE".

❹ Additional steps are needed to generate some more complicated variables, such as *RACE*. Since some subjects have multiple races, we should set *RACE*="MULTIPLE". At the same time, we need to split multiple races into separate sub *RACE* variables in the SUPPDM domain, such as *"RACE1"*= "AMERICAN INDIAN OR ALASKA AMERICAN", *RACE2*="ASIAN", *RACE3* ="BLACK OR AFRICAN AMERICAN", *RACE4*="NATIVE HAWAIIAN OR OTHER PACIFIC ISLANDERS", *RACE5*="WHITE", *RACE6*= "NOT REPORTED", *RACE7*= "UNKNOWN" given the *RACE* is "MULTIPLE" in the DM domain.

Note: We have created a second domain, called SUPPDM. While we do not go into detail about the supplemental (SUPP) domains, we provide an example above about when to use them. In general, SUPP domains are only created when there is information in the parent domain that doesn't "fit" in the parent domain per the SDTM IG.

❺ Based on the CRF and Specification, if *RACE*="MULTIPLE", then we need to map each race value from DM domain to SUPPDM. (As mentioned earlier, we only map data that doesn't "fit" into the parent domain into the child/ "SUPP" domain.) Given *RACE*= "MULTIPLE", and if race's value does not belong to one of the pre-specified values, then set *RACE*= "" in the DM domain and map *RACEOTH*="OTHER" in SUPPDM.

While DM and other domains have their own respective CRF pages, there will be times when we need to merge a domain (often DM) with other domain(s) in order to populate required SDTM variables. Next, we demonstrate how to populate one of the most important variables – treatment ARM.

```
/*Dropping records with the same ARMCD in order to merge back with DM
variables*/
/*Remove duplicate records with the same ARM)*/
proc sort data=RAW.TA out=TA(keep=armcd arm) nodupkey;
     by ARMCD;
run;
/*Merge DM1 with TA domain using ARMCD*/
proc sql;
     create table DM2 as select a.*,b.ARM length 200 from DM1 a left join
TA b on a.ARMCD=b.ARMCD;
❻
quit;

/*Create RFSTDTC and RFENDTC from EX domain*/

data EX1;
```

```
   set RAW.EX(keep=SUBJID EXSTDAT);    ❼
   EXDTS=datepart(EXSTDAT);
   EXTMS=timepart(EXSTDAT);
   EXDTS_DT=put(EXDTS,yymmdd10.);
   EXDTS_TM=put(EXTMS,time8.);
   EXSTDTC=strip(EXDTS_DT)||"T"||strip(EXDTS_TM);
run;

data EX2(rename=(EXSTDTC=RFSTDTC))
     EX3(rename=(EXSTDTC=RFENDTC));
   set EX1;
   by SUBJID EXSTDTC;
   if first.SUBJID then output EX2;
   if last.SUBJID then output EX3;
run;
proc sql;
    create table DM3 as select a.*,b.RFSTDTC from DM2 a left join EX2 b on
a.SUBJID=b.SUBJID;                              ❽
       create table DM4 as select a.*,b.RFENDTC from DM3 a left join EX3 b on
a.SUBJID=b.SUBJID;
quit;
```

❻ Merge DM and RAW.TA (Trial *ARM*) domain to populate treatment *ARM* variable *ARMCD* which we then use to populate *ARM*.

❼ Derive Reference Start Date and Time *(RFSTDTC)* from Date of First Exposure. *RFSTDTC* is derived from the variable *EXSTDAT* in RAW.EX data set. We use datepart() and timepart() functions to split the date and time components apart, then use put() function with yymmdd10. and put() with time8. to convert numeric value to character format. Finally, use concatenation || function to combine date and time parts with "T" to derive *EXSTDTC*. Finally, we use first.statement to capture the first record in the BY group.

Similarly, to derive Reference End Date and Time (*RFENDTC*) use last.statement to capture the last record in the BY group.

❽ Merge DM2 with EX2 to capture RFSTDTC variable from EX2 merge with DM2 domain based on *SUBJID*; then merge DM3 with EX3 to capture *RFENDTC* variable from EX3 and merge with DM3 domain.

```
data Final;
/*Defining DOMAIN, STUDYID, USUBJID, ACTARM, ACTARMCD*/
   set DM4;
   length ACTARMCD ARMCD $20. ARM ACTARM $200.;
   DOMAIN="DM";
   STUDYID="ABC-001";
   USUBJID=STRIP(STUDYID)||"-"||STRIP(SITEID)||"-"||STRIP(SUBJID);
   ACTARM=strip(ARM);
   ACTARMCD = ARMCD;
   format _all_;
   informat _all_;                    ❾
run;
```

❾ We use the statements format _all_ and informat _all_ to remove all formats and informats previously assigned in the data sets.

```
libname SDTM ".../directory";
data SDTM.DM(label="Demographics");
/*Assign variable attributes such as label and length to conform with SDTM.DM
Specification (these will also be the same attributes as the SDTM IG).*/
attrib
     STUDYID     label = "Study Identifier"                    length = $20
     DOMAIN      label = "Domain Abbreviation"                 length = $2
     USUBJID     label = "Unique Subject Identifier"           length = $40
     SUBJID      label = "Subject Identifier for the Study"    length = $20
     RFSTDTC     label = "Subject Reference Start Date/Time"   length = $20
     RFENDTC     label = "Subject Reference End Date/Time"     length = $20
     BRTHDTC     label = "Date/Time of Birth"                  length = $20
     SITEID      label = "Study Site Identifier"               length = $10
     AGE         label = "Age"                                 length = 8
     AGEU        label = "Age Units"                           length = $10
     SEX         label = "Sex"                                 length = $2
     RACE        label = "Race"                                length = $100
     ETHNIC      label = "Ethnicity"                           length = $60
     ARM         label = "Description of Planned Arm"          length = $200
     ARMCD       label = "Planned Arm Code"                    length = $20
     ACTARMCD    label = "Actual Arm Code"                     length = $20
     ACTARM      label = "Description of Actual Arm"           length = $200
     COUNTRY     label = "Country"                             length = $4
        ;
  set Final;
  keep STUDYID DOMAIN USUBJID SUBJID RFSTDTC RFENDTC BRTHDTC SITEID AGE AGEU
SEX RACE ETHNIC ARMCD ARM ACTARMCD ACTARM COUNTRY
        ;
run;
```

The final DM domain is saved in the SDTM library, as shown in Figure 3.4.

Figure 3.4: SDTM.DM

SITEID	AGE	AGEU	SEX	RACE	ETHNIC
001	20	YEARS	F	WHITE	NOT HISPANIC OR LATINO
001	20	YEARS	F	WHITE	NOT HISPANIC OR LATINO
001	30	YEARS	M	WHITE	NOT HISPANIC OR LATINO
001	35	YEARS	M	WHITE	NOT HISPANIC OR LATINO
001	22	YEARS	M	WHITE	NOT HISPANIC OR LATINO
001	34	YEARS	M	WHITE	NOT HISPANIC OR LATINO
001	24	YEARS	M	WHITE	NOT HISPANIC OR LATINO
001	20	YEARS	M	BLACK OR AFRICAN AMERICAN	NOT HISPANIC OR LATINO
001	30	YEARS	M	BLACK OR AFRICAN AMERICAN	NOT HISPANIC OR LATINO
001	30	YEARS	M	MULTIPLE	NOT HISPANIC OR LATINO

ARM	ARMCD	ACTARMCD	ACTARM	COUNTRY
DRUG A 10 mg	DRUG A	DRUG A	DRUG A 10 mg	USA
DRUG A 10 mg	DRUG A	DRUG A	DRUG A 10 mg	USA
DRUG A 10 mg	DRUG A	DRUG A	DRUG A 10 mg	USA
DRUG A 10 mg	DRUG A	DRUG A	DRUG A 10 mg	USA
DRUG A 10 mg	DRUG A	DRUG A	DRUG A 10 mg	USA
DRUG A 10 mg	DRUG A	DRUG A	DRUG A 10 mg	USA
DRUG A 10 mg	DRUG A	DRUG A	DRUG A 10 mg	USA
DRUG A 10 mg	DRUG A	DRUG A	DRUG A 10 mg	USA
DRUG A 10 mg	DRUG A	DRUG A	DRUG A 10 mg	USA
DRUG A 10 mg	DRUG A	DRUG A	DRUG A 10 mg	USA

SUPPDM Programming

In Figure 3.4, note that subject ABC-001-001-010 has multiple races. As such, we dig into the RAW.DM and see that ABC-001-001-010's *RACE* is "Black or African American" and "AMERICAN INDIAN OR ALASKA NATIVE". Since we can record only one *RACE* in DM, we will need to map all the subjects' races into SUPPDM.

```
data SUPPDM;
/*Create SUPPxx Variables which will always be QNAM, QLABEL, QVAL, QORIG,
IDVAR, IDVARVAL and RDOMAIN. */
  set Final;
  length RDOMAIN $2. IDVAR $8. QNAM USUBJID IDVARVAL QLABEL $40. QORIG QVAL
$100.;
  RDOMAIN="DM";
  IDVAR="";
  IDVARVAL="";
  QORIG="CRF";
  if RACE="MULTIPLE" and ^missing(RACE_AINDIAN) then do;
     QNAM="RACE1";
     QLABEL="American Indian/Alaska Native";
     QVAL="AMERICAN INDIAN OR ALASKA NATIVE";
   output;
end;
  if RACE="MULTIPLE" and ^missing(RACE_ASIAN) then do;
     QNAM="RACE2";
     QLABEL="ASIAN";
```

```
       QVAL="ASIAN";
    output;
  end;
  if RACE="MULTIPLE" and ^missing(RACE_BLACK) then do;
      QNAM="RACE3";
      QLABEL="Black or African American";
      QVAL="BLACK OR AFRICAN AMERICAN";
      output;
  end;
  if RACE="MULTIPLE" and ^missing(RACE_HAWAIIAN) then do;
      QNAM="RACE4";
      QLABEL="Native Hawaiian/Pacific Islander";
      QVAL="NATIVE HAWAIIAN OR OTHER PACIFIC ISLANDER";
      output;
  end;
  if RACE="MULTIPLE" and ^missing(RACE_WHITE) then do;
      QNAM="RACE5";
      QLABEL="White";
      QVAL="WHITE";
      output;
  end;
  if RACE="MULTIPLE" and ^missing(RACE_NOREPORT) then do;
      QNAM="RACE6";
      QLABEL="Not reported";
      QVAL="NOT REPORTED";
      output;
  end;
  if RACE="MULTIPLE" and ^missing(RACE_UNKNOWN) then do;
      QNAM="RACE7";
      QLABEL="Unknown";
      QVAL="UNKNOWN";
    output;
  end;
  if RACE="MULTIPLE" and ^missing(RACE_OTHER) then do;
      QNAM="RACE8";
      QLABEL="Other";
      QVAL="OTHER";
      output;
  end;
    QORIG="CRF";

run;
```

As we describe above, if *RACE*="MULTIPLE" and either one of the *RACE* variables is not missing, we need to split multiple races into individual *RACE* variables in the SUPPDM domain, such as *QVAL*= "AMERICAN INDIAN OR ALASKA AMERICAN" if *QNAM*= "RACE1", *QVAL* = "ASIAN" if *QNAM*= "RACE2", *QVAL* = "BLACK OR AFRICAN AMERICAN" if *QNAM* = "RACE3", *QVAL* = "NATIVE HAWAIIAN OR OTHER PACIFIC ISLANDERS"if *QNAM* = "RACE4", *QVAL*= "WHITE" if *QNAM* = "RACE5", *QVAL* = "NOT REPORTED" if *QNAM* = "RACE6", *QNAM*= "UNKNOWN" if *QVAL* = "RACE7", given "RACE is MULTIPLE" in the DM domain.

Note: For readers familiar with the SAS Macro language, it's obvious that it would be more efficient to code using a SAS Macro. However, for the purpose of this book, we want to keep the code as readable as possible.

```
libname SDTM ".../directory";
data SDTM.SUPPDM(label="Supplemental Qualifiers for DM");
/*Assign variable attributes such as label and length to conform with SDTM.
SUPPDM Specification (these will also be the same attributes as the SDTM
IG).*/

attrib
      STUDYID        label = "Study Identifier"               length = $20
      RDOMAIN        label = "Related Domain Abbreviation"    length = $2
      USUBJID        label = "Unique Subject Identifier"      length = $40
      IDVAR          label = "Identifying Variable"           length = $8
      IDVARVAL       label = "Identifying Variable Value"     length = $40
      QNAM           label = "Qualifier Variable Name"        length = $40
      QLABEL         label = "Qualifier Variable Label"       length = $40
      QVAL           label = "Data Value"                     length = $100
      QORIG          label = "Origin"                         length = $100
      ;
  set SUPPDM ;
   keep STUDYID RDOMAIN USUBJID IDVAR IDVARVAL QNAM QLABEL QVAL QORIG
      ;
run;
```

The final SUPPDM data set is saved in the SDTM library, as shown in Figure 3.5.

Figure 3.5: SDTM.SUPPDM

	STUDYID	RDOMAIN	USUBJID	IDVAR	IDVARVAL
1	ABC-001	DM	ABC-001-001-010		
2	ABC-001	DM	ABC-001-001-010		

QNAM	QLABEL	QVAL	QORIG
RACE1	American Indian/Alaska Native	AMERICAN INDIAN OR ALASKA NATIVE	CRF
RACE3	Black or African American	BLACK OR AFRICAN AMERICAN	CRF

3.6.2 Disposition (DS)

The DS domain is also very important because it provides a summary of clinical events from many different raw data sets. For example, the DS domain includes information about all subjects who entered the study and protocol milestones, such as randomization date, subjects' completion status and date and reason/dates for discontinuation of treatment or study.

Structure of DS

The structure of DS consists of Identifier, Timing, Topic, and Qualifier variables, as follows.

Identifier Variables:

- STUDYID: Required variable in character format, the unique study identifier. Each study has a unique study ID. In the example study described in this book, we populate it with "ABC-001".
- DOMAIN: Required variable in character format, compliant with SDTM Implementation Guide. *DOMAIN* sets to "DS".
- USUBJID: Required variable in character format used to uniquely identify a subject across all studies. The *USUBJID* variable is derived using: *STUDYID- SITED-SUBJID*, such as "ABC-001-001-001".
- DSSEQ: Required variable in numeric format, the sequence number given to ensure uniqueness of subject records within a domain.

Timing Variables:

- DSSTDTC: Expected variable in character format, the description of the start date and time of Disposition Event.

Topic Variables:

- DSTERM: Required variable in character format, the reported term for the Disposition Event.

Qualifier Variables:

- DSDECOD: Required variable in character format, the standardization Disposition Term.
- DSCAT: Expected variable in character format, the category for Disposition Event.

DS eCRF Reference

The Disposition information is collected in each stage of a clinical trial, including "Informed Consent", "Treatment", and "Follow-Up". In practice, and as mentioned earlier, Disposition information is recorded in multiple CRF pages. In our simulated study, the CRF from CDSIC only populates the subject's disposition at the End of Study, as shown in Figure 3.6. In the eCRF End of Study page, if the subject completes the study, then *DSCAT*= "DISPOSITION EVENT",

Figure 3.6: SDTM Annotated Disposition (DS)

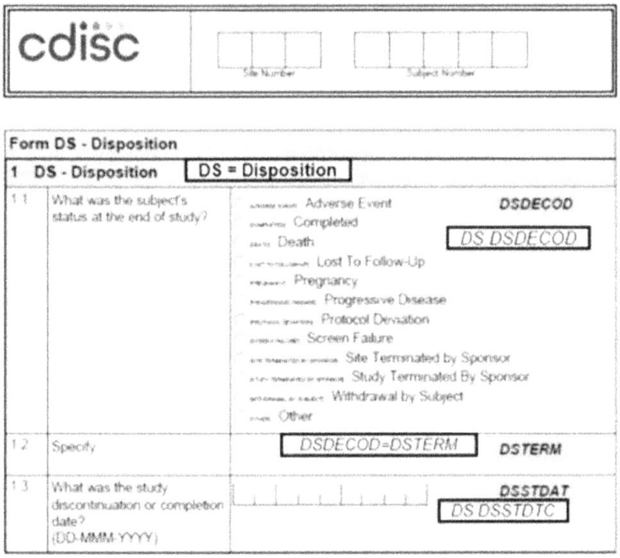

cdisc

Site Number | Subject Number

Form DS - Disposition

1 DS - Disposition | DS = Disposition

1.1	What was the subject's status at the end of study?	Adverse Event Completed Death Lost To Follow-Up Pregnancy Progressive Disease Protocol Deviation Screen Failure Site Terminated by Sponsor Study Terminated By Sponsor Withdrawal by Subject Other DSDECOD DS.DSDECOD
1.2	Specify	DSDECOD=DSTERM DSTERM
1.3	What was the study discontinuation or completion date? (DD-MMM-YYYY)	DSSTDAT DS.DSSTDTC

DSDECOD=DSTERM= "COMPLETED". If the subject does not complete the study, then the eCRF lists the reason of the discontinuation, for example, due to "Adverse Event", "Death", "Lost to Follow-Up", "Pregnancy", "Screen Failure", etc. In the case where the subject does not complete the study, then *DSCAT*= "DISPOSITION EVENT", *DSDECOD=DSTERM*= "ADVERSE EVENT", "DEATH", etc.

In order to be consistent with the CDISC eCRF, we provide annotations for the Disposition eCRF page and show readers the process of creating the DS domain.

DS Specification

Table 3.3: DS Specification

Variable	Label	Type	Length	Control Terminology	Origin	Notes
STUDYID	Study Identifier	char	20		Protocol: Set to "ABC-001"	This information that programmer can find in multiple raw data sets, such as RAW.DM;
DOMAIN	Domain Abbreviation	char	2	DOMAIN	Assigned: Set to "DS"	

(*Continued*)

Table 3.3: (*Continued*)

Variable	Label	Type	Length	Control Terminology	Origin	Notes
USUBJID	Unique Subject Identifier	char	40		Derived: Concatenate among three variables below: STUDYID-SITED-SUBJID	
DSSEQ	Sequence Number	char	20		Derived: Sorted by USUBJID, DSSTDTC, DSDECOD. Start with value 1 for the first row for each subject and increment by 1 at each successive record.	
DSTERM	Reported Term for the Disposition Event	char	200		CRF	Map the uppercase value in terms of difference scenarios on CRF.
DSDECOD	Standardized Disposition Term	char	200		CRF	The value is equal to DSTERM, refer to CRF.
DSCAT	Category for Disposition Event	char	40		Assigned: Set to DISPOSITION EVENT at the End of Study	
DSSTDTC	Start Date/ Time of Disposition Event	char	20		CRF	

DS Programming

In order to create SDTM.DS, we first run a PROC CONTENTS on the raw data. Please see the code below.

Note: For every domain, the corresponding SAS name is domain.sas. For example, DS is created by ds.sas, so all code in this section belongs to ds.sas.

```
/*Begin writing SAS program ds.sas*/
/*show structure of the raw DS dataset*/
proc contents data=RAW.DS;
run;
```

First, as seen in Figure 3.7, we use PROC CONTENTS to see that there are 9 variables in the raw DS data set. Further, we can see the Type, Length (Len), Format, and Informat of each variable. These variable attributes are assigned by SAS based on the raw data format. The Label is assigned in SAS based on the first row of the raw DM data set in the Excel sheet.

Note: The purpose of running PROC CONTENTS is to get a clear idea of what the raw metadata looks like and compare how similar it is to the SDTM domain that we will create from it.

```
/*Create the 1st set of DS variables using existing variables from RAW.DS*/
/*termination- end of dosing*/
data DS1;
  set RAW.DS;
  /*Define DOMAIN, STUDYID, SITEID, USUBJID, DSSTDTC*/
  DOMAIN="DS";
  STUDYID="ABC-001";
  SITEID=SITE;
  USUBJID=STRIP(STUDYID)||"-"||STRIP(SITEID)||"-"||STRIP(SUBJID);
  DSCAT="DISPOSITION EVENT";
  DSSTDTC=strip(put(DSSTDAT,yymmdd10.));    ❶

  /*Derive DSTERM*/
  if upcase(DSDECOD)="COMPLETED" then DSTERM="COMPLETED";
  else if upcase(DSDECOD)="ADVERSE EVENT" then DSTERM="ADVERSE EVENT";
  else if upcase(DSDECOD)="DEATH" then DSTERM="DEATH";
  else if upcase(DSDECOD)="Lost To Follow-Up" then DSTERM="Lost To
Follow-Up";
else if upcase(DSDECOD)="PREGANCY" then DSTERM="PREGANCY";
  else if upcase(DSDECOD)="PROGRESSIVE DISEASE" then DSTERM="PROGRESSIVE
DISEASE";
  else if upcase(DSDECOD)="PROTOCOL DEVIATION" then DSTERM="PROTOCOL
DEVIATION";
  else if upcase(DSDECOD)="SCREEN FAILURE" then DSTERM="SCREEN FAILURE";
  else if upcase(DSDECOD)="SITE TERMINATED BY SPONSOR" then DSTERM="SITE
TERMINATED BY SPONSOR";
```

Figure 3.7: Alphabetic List of Variables and Attributes

Alphabetic List of Variables and Attributes

#	Variable	Type	Len	Format	Informat	Label
2	COUNTRY	Char	13	$13	$13	COUNTRY
8	DSDECOD	Char	9	$9	$9	DSDECOD
9	DSDISC	Char	1	$1	$1	DSDISC
6	DSSTDAT	Num	8	MMDDYY10		DSSTDAT
5	EVENT	Char	12	$12	$12	EVENT
7	FORM	Char	20	$20	$20	FORM
3	SITE	Char	3	$3	$3	SITE
1	STUDYID	Char	7	$7	$7	STUDYID
4	SUBJID	Char	3	$3	$3	SUBJID

```
  else if upcase(DSDECOD)="STUDY TERMINATED BY SPONSOR" then DSTERM="STUDY
TERMINATED BY SPONSOR";
  else if upcase(DSDECOD)="WITHDRAWN BY SUBJECT" then DSTERM="WITHDRAWN BY
SUBJECT";
  else if upcase(DSDECOD)="OTHER" then DSTERM="OTHER";
run;
```

❶ Be diligent when you create variables with date and time, such as *DSSTDTC*, which is derived from the variable DSSTADT (numeric) in the RAW.DS data set. We use put() with format *yymmdd10.* to convert numeric value to character in *yyyy-mm-dd.* format.

```
/*Sort data set DS1 by USUBJID, DSSTDTC, DSDECOD*/
proc sort data=DS1 out=DS2;
  by USUBJID DSSTDTC DSDECOD;
run;
/*Derive DSSEQ*/
data Final;
  set DS2;
  length DSSEQ 8.;            ❷
  by USUBJID DSSTDTC DSDECOD;
  if FIRST.USUBJID then DSSEQ=0;
    DSSEQ+1;
  output;
  format _all_;
  informat _all_;             ❸
run;
```

❷ Create the *DSSEQ* variable, which is used to uniquely record the sequence number of a subject's disposition events within the DS domain. In order to create *DSSEQ*, we sort by *USUBJID, DSSTDTC,* and *DSDECOD*, with the first record in the BY group assigned 0, so *DSSEQ*=0 and in the next iteration the value of *DSSEQ* is *DSSEQ*+1 (0+1=1), and the new value of 1 for

DSSEQ will override the former value of 0. The *DSSEQ* value is written in the first output until it encounters the OUTPUT statement. Then SAS goes back to the *DSSEQ*+1 statement and increments by 1 for each subsequent record. For example, we listed two subjects in Figure 3.8 below to help readers understand how SAS counts the SEQ variable in the SAS Program Data Vector (PDV).

❸ The statements format _all_ and informat _all_ remove all formats and informats previously assigned in the data sets.

The final DS domain is saved in the SDTM library, as shown in Figure 3.9.

```
libname SDTM ".../directory";
data SDTM.DS (label="Disposition");
/*Assign variable attributes such as label and length to conform with SDTM.DS
Specification (these will also be the same attributes as the SDTM IG).*/
attrib
  STUDYID       label = "Study Identifier"                       length = $20
  DOMAIN        label = "Domain Abbreviation"                    length = $2
  USUBJID       label = "Unique Subject Identifier"              length = $40
  DSSEQ         label = "Sequence Number"                        length = 8
  DSTERM        label = "Reported Term for the Disposition Event" length = $200
  DSDECOD       label = "Standardized Disposition Term"          length = $200
  DSCAT         label = "Category for Disposition Event"         length = $40
  DSSTDTC       label = "Start Date/Time of Disposition Event"   length = $20
  ;
  set Final;
    keep STUDYID DOMAIN USUBJID DSSEQ DSTERM DSDECOD DSCAT DSSTDTC
  ;
run;
```

The final DS domain output is saved in the SDTM library, as shown in Figure 3.9.

Figure 3.8: SAS Program Data Vector

Figure 3.9: SDTM.DS

	STUDYID	DOMAIN	USUBJID	DSSEQ	DSTERM
1	ABC-001	DS	ABC-001-001-001	1	COMPLETED
2	ABC-001	DS	ABC-001-001-002	1	COMPLETED
3	ABC-001	DS	ABC-001-001-003	1	COMPLETED
4	ABC-001	DS	ABC-001-001-004	1	COMPLETED
5	ABC-001	DS	ABC-001-001-005	1	COMPLETED
6	ABC-001	DS	ABC-001-001-006	1	COMPLETED
7	ABC-001	DS	ABC-001-001-007	1	COMPLETED
8	ABC-001	DS	ABC-001-001-008	1	COMPLETED
9	ABC-001	DS	ABC-001-001-009	1	COMPLETED
10	ABC-001	DS	ABC-001-001-010	1	COMPLETED

DSDECOD	DSCAT	DSSTDTC
COMPLETED	DISPOSITION EVENT	2021-01-04
COMPLETED	DISPOSITION EVENT	2021-01-04
COMPLETED	DISPOSITION EVENT	2021-01-04
COMPLETED	DISPOSITION EVENT	2021-01-04
COMPLETED	DISPOSITION EVENT	2021-01-04
COMPLETED	DISPOSITION EVENT	2021-01-04
COMPLETED	DISPOSITION EVENT	2021-01-04
COMPLETED	DISPOSITION EVENT	2021-01-04
COMPLETED	DISPOSITION EVENT	2021-01-04
COMPLETED	DISPOSITION EVENT	2021-01-04

3.6.3 Adverse Events (AE)

The Adverse Events (AE) domain contains data that describes untoward medical occurrences in subjects or subjects who are administered a pharmaceutical product, which might not necessarily have a causal relationship with the treatment.

Structure of AE

The structure of AE consists of Identifier, Timing, Topic, and Qualifier variables, as follows.

Identifier Variables:

- STUDYID: Required variable in character format, the unique study identifier. Each study has a unique study ID. In the example study described in this book, we populate it with "ABC-001".
- DOMAIN: Required variable in character format, compliant with the SDTM Implementation Guide DOMAIN sets to "DS".

- USUBJID: Required variable in character format used to uniquely identify a subject across all studies. The USUBJID variable is derived using: STUDYID- SITED-SUBJID, such as "ABC-001-001-001".
- AESEQ: Required variable in numeric format, the sequence number given to ensure uniqueness of subject records within a domain.

Timing Variables:

- AESTDTC: Expected variable in character format, the description of the start date and time of Adverse Event, representing in ISO 8601 character format.
- AEENDTC: Expected variable in character format, the description of the end date and time of Adverse Event, representing in ISO 8601 character format.

Topic Variables:

- AETERM: Required variable in character format, the reported term for the Adverse Events.

Qualifier Variables:

- AELLT: Expected variable in character format, the dictionary-derived text description of the lowest level term.
- AELLTCD: Expected variable in character format, the dictionary-derived code for the lowest level term.
- AEDECOD: Required variable in character format, the dictionary-derived text description of reported term (AETERM), equivalent to the Preferred Term (PT) variable.
- AEPTCD: Expected variable in character format, the description of the dictionary-derived code for the Preferred Term.
- AEHLT: Expected variable in character format, the dictionary-derived text description of the high-level term for the primary System Organ Class.
- AEHLTCD: Expected variable in character format, the dictionary-derived code for the high-level term for the primary System Organ Class.
- AEHLGT: Expected variable in character format, the dictionary-derived text description of the High-Level Group Term for the primary System Organ Class.
- AEHLGTCD: Expected variable in character format, the dictionary-derived code for the High-Level Group Term for the primary System Organ Class.
- AEBODSYS: Expected variable in character format, the Body System or Organ Class used by the sponsor from the coding dictionary, such as *MedDRA*.
- AEBODYSCD: Expected variable in character format, the code for the Body System or Organ Cass used by sponsor.
- AESOC: Expected variable in character format, the dictionary-derived text description of the primary System Organ Class (SOC), equivalent to *AEBODSYS*, given the primary SOC was used for analysis.
- AESER: Expected variable in character format, "Is this a serious event?" The values are "Yes" or "No", "Y" and "N", respectively.

- AEACN: Expected variable in character format, the changes on the study treatment as a result of the event. Valid values contain: "Dose Increased", "Dose Not Changed", "Dose Rate Reduced", "Dose Reduced", "Drug Interrupted", "Drug Withdrawn", "Not Applicable", or "Unknown".
- AEREL: Expected variable in character format, the record of the investigator's opinion as to the causality of the event to the treatment.
- AESMIE: Permissible variable in character format, the description of the adverse event, which is a medically important event not covered by another serious event.
- AEOUT: Permissible variable in character format, the description of the outcome of an event.
- AESCONG: Permissible variable in character format, the congenial anomaly or birth defect.
- AESDISAB: Permissible variable in character format, the persist or significant disability.
- AESDTH: Permissible variable in character, the format results in death.
- AESHOSP: Permissible variable in character format, "Did the serious event require or prolong hospitalization?" The values are "Yes" or "No", "Y" and "N", respectively.
- AESLIFE: Permissible variable in character format, "Is life threatening?" The values are "Yes" or "No", "Y" and "N", respectively.

Note: The variables from *AELLT* to *AESOC* are not collected on the CRF. These are "coding variables". In this AE example, the actual event is entered in plain English on the CRF, and these terms are coded consistently using a dictionary used throughout the industry.

AE eCRF Reference

Adverse Events information is collected in each stage including "Screening", "Treatment", and "Follow-Up". In our simulated study, the eCRF from CDISC is used as an example to create our AE domain.

Figure 3.10: SDTM Annotated Adverse Events

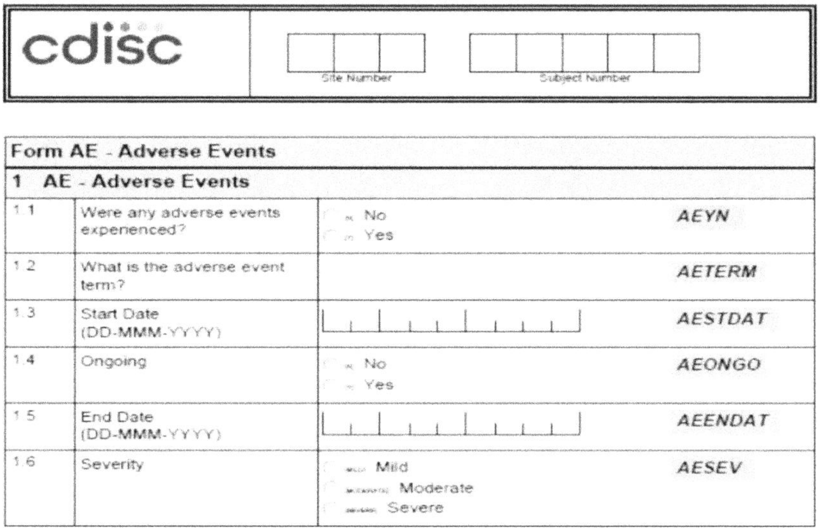

Site Number | Subject Number

Form AE - Adverse Events			
1	**AE - Adverse Events**		
1.1	Were any adverse events experienced?	No Yes	*AEYN*
1.2	What is the adverse event term?		*AETERM*
1.3	Start Date (DD-MMM-YYYY)		*AESTDAT*
1.4	Ongoing	No Yes	*AEONGO*
1.5	End Date (DD-MMM-YYYY)		*AEENDAT*
1.6	Severity	Mild Moderate Severe	*AESEV*

1.7	Was the adverse event serious?	☐ No **AESER** ☐ Yes Did the adverse event result in death? ☐ No **AESDTH** ☐ Yes Was the adverse event life threatening? ☐ No **AESLIFE** ☐ Yes Did the adverse event result in initial or prolonged hospitalization for the subject? ☐ No **AESHOSP** ☐ Yes Did the adverse event result in disability or permanent damage? ☐ No **AESDISAB** ☐ Yes Was the adverse event associated with a congenital anomaly or birth defect? ☐ No **AESCONG** ☐ Yes Was the adverse event a medically important event not covered by other serious criteria? ☐ No **AESMIE** ☐ Yes
1.8	Relationship to Study Treatment	☐ No **AEREL** ☐ Yes

1.9	Action Taken with Study Treatment	☐ Dose Increased **AEACN** ☐ Dose Not Changed ☐ Dose Rate Reduced ☐ Dose Reduced ☐ Drug Interrupted ☐ Drug Withdrawn ☐ Not Applicable ☐ Unknown
1.10	Other Action Taken	**AEACNOTH**
1.11	Outcome	☐ Fatal **AEOUT** ☐ Not Recovered or Not Resolved ☐ Recovered or Resolved ☐ Recovered or Resolved with Sequelae ☐ Recovering or Resolving ☐ Unknown

AE Specification

Table 3.4: AE Specification

Variable	Label	Type	Length	Control Terminology	Origin	Notes
STUDYID	Study Identifier	char	20		Protocol: Set to "ABC-001"	This information that programmer can find in multiple raw data sets, such as RAW. DM

(Continued)

Table 3.4: (*Continued*)

Variable	Label	Type	Length	Control Terminology	Origin	Notes
DOMAIN	Domain Abbreviation	char	2	DOMAIN	Assigned: Set to "AE"	
USUBJID	Unique Subject Identifier	char	40		Derived: Concatenate among three variables below: STUDYID- SITED- SUBJID	
AESEQ	Sequence number	num	8		Derived: Sequence number given to ensure uniqueness of subject records	Sort by USUBJID, AESTDTC, AEENDTC, AETERM, starts with value 1 for the first row of each subject, then increment 1 at each successive record for each subject.
AETERM	Reported Term for the Adverse Event	char	200		CRF	Raw data set: AE.AETERM
AELLT	Lowest Level Term	char	100	MEDDRA	Assigned: Assigned per MedDRA dictionary.	Raw data set: AE.LLT
AELLTCD	Lowest Level Term Code	num	8	MEDDRA	Assigned: Assigned per MedDRA dictionary.	Raw data set: AE.LLTCD
AEDECOD	Dictionary-Derived Term	char	200	MEDDRA	Assigned: Assigned per MedDRA dictionary.	Raw data set: AE.PT

(*Continued*)

Table 3.4: (*Continued*)

Variable	Label	Type	Length	Control Terminology	Origin	Notes
AEPTCD	Preferred Term Code	num	8	MEDDRA	Assigned: Assigned per MedDRA dictionary.	Raw data set: AE.PT_CD
AEHLT	High Level Term	char	200	MEDDRA	Assigned: Assigned per MedDRA dictionary.	Raw data set: AE.HLT
AEHLTCD	High Level Term Code	num	8	MEDDRA	Assigned: Assigned per MedDRA dictionary.	Raw data set: AE.HLTCD
AEHLGT	High Level Group Term	char	200	MEDDRA	Assigned: Assigned per MedDRA dictionary.	Raw data set: AE.HLGT
AEHLGTCD	High Level Group Term Code	num	8	MEDDRA	Assigned: Assigned per MedDRA dictionary.	Raw data set: AE.HLGTCD
AEBODSYS	Body System or Organ Class	char	20		Assigned: Assigned per MedDRA dictionary	Raw data set: AE.SOC
AESER	Serious Event	char	2		CRF	Raw data set: AE.AESER_YN
AEACN	Action Taken with Study Treatment	char	50		CRF	Raw data set: AE.AEACN
AEREL	Causality	char	50		CRF	Raw data set: AE.AEREL
AEOUT	Outcome of Adverse Event	char	50		CRF	Raw data set: AE.AEOUT
AESCONG	Congenital Anomaly or Birth Defect	char	2		CRF	Set to "Y" when raw data set AE.AESCONG is not null.

(*Continued*)

Table 3.4: (*Continued*)

Variable	Label	Type	Length	Control Terminology	Origin	Notes
AESDISAB	Persist or Significant Disability	char	2		CRF	Set to "Y" when raw data set AE.AESDISAB is not null.
AESDTH	Results in Death	char	2		CRF	Set to "Y" when raw data set AE.AESDTH is not null.
AESHOSP	Requires or Prolongs Hospitalization	char	2		CRF	Set to "Y" when raw data set AE.AESHOSP is not null.
AECONTRT	Concomitant or Additional Treatment Given	char	2		CRF	Set to "Y" when raw data set AE.AECONTRT is "Yes'; otherwise, set to "N" when AECONTRT is "No"
AETOXGR	Standard Toxicity Grade	char	2		CRF Set to "1" when AESEV="MILD". Set to "2" when AESEV="MODER-ATE". Set to "3" when AESEV="SEVERE". Set to "4" when AESEV="LIFE THREATHEING"	Raw data set, AR.AESEV, uppercase values
AESTDTC	Start Date/ Time of Adverse Event	char	20		CRF	Raw data set: AE.AESDTA

Table 3.4: (*Continued*)

Variable	Label	Type	Length	Control Terminology	Origin	Notes
AEENDTC	End Date/Time of Adverse Event	char	20		CRF	Raw data set: AE.AEENDAT
AESTDY	Study Day of Start of Adverse Event	num	8		Derived: Difference between AE.AESTDTC and DM.RFSDTDC	
AEENDY	Study Day of End of Adverse Event	num	8		Derived: Difference between AE.AEENDTC and DM.RFSDTDC	
AEENRTPT	End Relative to Reference Time Point	char	40		Assigned: Set to "ONGOING" when AEENDTC is Null and AEOUT = "NOT RECOVERED/ NOT RESOLVED". Set to "UNKNOWN" when AEENDTC is Null and AEOUT = "UNKNOWN".	
AEENTPT	End Reference Time Point	char	40		Assigned: Set to "END OF SAFETY FOLLOW-UP" when AEENRTPT is either "ONGOING" or "UNKNOWN". Otherwise, set to null.	

AE Programming

In order to create SDTM.AE, we first run a PROC CONTENTS on the raw data. Please see the code below.

Note: For every domain, the corresponding SAS name is domain.sas. For example, AE is created by ae.sas, so all code in this section belongs to ae.sas.

```
/*Begin writing SAS program ae.sas*/
/*show structure of the raw AE dataset*/
proc contents data=RAW.AE;
run;
```

First, as seen in Figure 3.11, we use PROC CONTENTS to see that there are 34 variables in the raw AE data set. Further, we can see the Type, Length (Len), Format, and Informat of each variable. These variable attributes are assigned by SAS based on the raw data format. The Label is assigned in SAS based on the first row of the raw AE data set in the Excel sheet.

Note: The purpose of running PROC CONTENTS is to get a clear idea of what the raw metadata looks like and compare how similar it is to the SDTM domain that we will create from it.

Figure 3.11: Alphabetic List of Variables and Attributes

Alphabetic List of Variables and Attributes

#	Variable	Type	Len	Format	Informat	Label
20	AEACN	Char	17	$17	$17	AEACN
19	AEENDAT	Num	8	MMDDYY10		AEENDAT
23	AEOUT	Char	21	$21	$21	AEOUT
21	AEREL	Char	3	$3	$3	AEREL
17	AESCONG	Char	1	$1	$1	AESCONG
18	AESDAT	Num	8	MMDDYY10		AESDAT
16	AESDISAB	Char	1	$1	$1	AESDISAB
14	AESDTH	Char	1	$1	$1	AESDTH
11	AESER	Char	2	$2	$2	AESER
22	AESEV	Char	9	$9	$9	AESEV
13	AESHOSP	Char	1	$1	$1	AESHOSP
15	AESLIFE	Char	1	$1	$1	AESLIFE
12	AESMIE	Char	2	$2	$2	AESMIE
10	AETERM	Char	5	$5	$5	AETERM
9	AEYN	Char	1	$1	$1	AEYN
2	COUNTRY	Char	13	$13	$13	COUNTRY
33	DICTYPE	Char	6	$6	$6	DICTYPE
5	[illegible]	Num	8	BEST		[illegible]
6	FORM	Char	14	$14	$14	FORM
7	FSEQ	Num	8	BEST		FSEQ
31	HLGT	Char	51	$51	$51	HLGT
32	HLGTCD	Num	8	BEST		HLGTCD
28	HLT	Char	48	$48	$48	HLT
29	HLTCD	Num	8	BEST		HLTCD
8	IGSEQ	Num	8	BEST		IGSEQ
26	LLT	Char	9	$9	$9	LLT
27	LLTCD	Num	8	BEST		LLTCD
34	[illegible]	Char	11	$11	$11	[illegible]
25	PT	Char	9	$9	$9	PT
30	PTCD	Num	8	BEST		PTCD
3	SITEID	Char	3	$3	$3	SITEID
24	SOC	Char	47	$47	$47	SOC
1	STUDYID	Char	7	$7	$7	STUDYID
4	SUBJID	Char	3	$3	$3	SUBJID

We use the PROC FORMAT value statement to create a format that converts the character strings on the left side to the corresponding values on the right side. For readers unfamiliar with CDISC standards, using this format helps us map a corresponding character variable, *AETOXGR,* from the variable *AESEV*, which has values of "Mild", "Moderate", and "Severe".

```
/*Assign the character value from "Mild", "Moderate" and "Severe" to
character value "1", "2" and "3" for variable AETOXGR with PROC FORMAT*/

proc format;
 value $AETOXGR
 "Mild"="1"
 "Moderate"="2"
 "Severe"="3"
;
quit;

data AE1;
/*Specify length for standard variables*/
  length STUDYID AESTDTC AEENDTC AEBODSYS $20      ❶
       DOMAIN AESER AESCONG AESDISAB AESDTH AESHOSP AESMIE AETOXGR $2
       USUBJID AEENRTPT AEENTPT $40
       AELLTCD AEPTCD AEHLTCD AEHLGTCD 8
       AETERM AEDECOD AEHLT AEHLGT $200
       AELLT $100
       AEACN AEREL AEOUT $50;
  /*Rename AETERM, AEACN, AESER, AEREL, AEOUT, AESCONG, AESDISAB, AESDTH,
AESHOSP, AESMIE */
  set RAW.AE(rename=(AETERM=AETERM_ AEACN=AEACN_ AESER=AESER_ AEREL=AEREL_
AEOUT=AEOUT_ AESCONG=AESCONG_ AESDISAB=AESDISAB_ AESDTH=AESDTH_
AESHOSP=AESHOSP_ AESMIE=AESMIE_));
  DOMAIN="DM";
  STUDYID="ABC-001";
  USUBJID=STRIP(STUDYID)||"-"||STRIP(SITEID)||"-"||STRIP(SUBJID);    ❷
  /*Define DOMAIN, STUDYID, USUBJID*/
  DOMAIN="AE";
  STUDYID="ABC-001";
  USUBJID=STRIP(STUDYID)||"-"||STRIP(SITEID)||"-"||STRIP(SUBJID);/*Derive
AETERM, AELLT, AELLTCD, AEDECOD, AEPTCD, AEHLT, AEHLTCD, AEHLGT, AEHLGTCD,
AEBODSYS, AEACN, AEOUT*/
  AETERM=strip(upcase(AETERM_));
  AELLT=strip(upcase(LLT));
  AELLTCD=LLTCD;
  AEDECOD=strip(upcase(PT));
  AEPTCD=PT_CD;
  AEHLT=strip(upcase(HLT));
  AEHLTCD=HLTCD;
  AEHLGT=strip(upcase(HLGT));
  AEHLGTCD=HLGTCD;
  AEBODSYS=strip(upcase(SOC));
  AEACN=strip(upcase(AEACN_));
  AEOUT=strip(upcase(AEOUT_));
```

```
/*Derive AESER, AESCONG, AESDISAB, AESDTH, AESHOSP, AESMIE */
if AESER_="Yes" then AESER="Y";
else if AESER_="No" then AESER="N";

if AEREL_="Yes" then AEREL="Y";
else if AEREL_="No" then AEREL="N";

if AESCONG_="Yes" then AESCONG="Y";
else if AESCONG_="No" then AESCONG="N";

if AESDISAB_="Yes" then AESDISAB="Y";
else if AESDISAB_="No" then AESDISAB="N";

if AESDTH_="Yes" then AESDTH="Y";
else if AESDTH_="No" then AESDTH="N";

if AESHOSP_="Yes" then AESHOSP="Y";
else if AESHOSP_="No" then AESHOSP="N";

if AESMIE_="Yes" then AESMIE="Y";
else if AESMIE_="No" then AESMIE="N";
/*Format AETOXGR, AESTDTC, AEENDTC */
AETOXGR=put(AESEV, AETOXGR.); ❸
AESTDTC=put(AESDAT,yymmdd10.);
AEENDTC=put(AEENDAT,yymmdd10.);
/*Derive AEENRTPT*/
if missing(AEENDTC) and AEOUT = "NOT RECOVERED OR NOT RESOLVED" then
AEENRTPT = "ONGOING";

else if missing(AEENDTC) and AEOUT = "UNKNOWN" then AEENRTPT = "UNKNOWN";
if AEENRTPT in("ONGOING", "UNKNOWN") then AEENTPT = "END OF STUDY";

  keep STUDYID AESTDTC AEENDTC DOMAIN AESER AESCONG AESDISAB AESDTH AESHOSP
AEBODSYS USUBJID AEENRTPT AEENTPT AELLTCD AEPTCD AEHLTCD AEHLGTCD AETERM
AEDECOD AEHLT AEHLGT AELLT AEACN AEREL AEOUT AETOXGR AESMIE;
run;
```

❶ Define the variables' lengths.
❷ Rename any variables from RAW.AE to re-assign variable attributes consistent with SDTM.
   AE variables.
❸ Most of variables' derivation in SDTM.AE are straightforward in the SDTM.AE specification,
   and each value should be uppercase. There are some exceptions, which are a little more complicated, such as using the put() function to convert formats from numeric to character, like
   AESDAT->AESTDTC and AEENDAT->AEENDTC.

```
/*Sort data set AE1 by USUBJID, AESTDTC, AEENDTC, AETERM */
proc sort data=AE1 out=AE2;
   by USUBJID AESTDTC AEENDTC AETERM;
 run;

data FINAL;
  set AE2;
  /*Derive AESEQ*/
  by USUBJID AESTDTC AEENDTC AETERM;
```

```
  if FIRST.USUBJID then AESEQ=0;      ❹
    AESEQ+1;
   output;
 format _all_;
 informat _all_;
run;
```

❹ Create the *AESEQ* variable, which is used to uniquely record the sequence number of a subject within a domain. To create *AESEQ*, please refer to the detailed explanation for *DSSEQ* variable under DS domain in Section 3.6.2.

```
libname SDTM"…/directory";
data SDTM.AE(label="Adverse Events");
/*Assign variable attributes such as label and length to conform with SDTM.AE
Specification (these will also be the same attributes as the SDTM IG).*/

attrib
  STUDYID      label = "Study Identifier"                       length = $20
  DOMAIN       label = "Domain Abbreviation"                    length = $2
  USUBJID      label = "Unique Subject Identifier"              length = $40
  AESEQ        label = "Sequence Number"                        length = 8
  AETERM       label = "Reported Term for the Adverse Event"    length = $200
  AELLT        label = "Lowest Level Term"                      length = $100
  AELLTCD      label = "Lowest Level Term Code"                 length = 8
  AEDECOD      label = "Dictionary-Derived Term"                length = $200
  AEPTCD       label = "Preferred Term Code"                    length = 8
  AEHLT        label = "High-Level Term"                        length = $200
  AEHLTCD      label = "High-Level Term Code"                   length = 8
  AEHLGT       label = "High-Level Group Term"                  length = $200
  AEHLGTCD     label = "High-Level Group Term Code"             length = 8
  AEBODSYS     label = "Body System or Organ Class"             length = $20
  AESER        label = "Serious Event"                          length = $2
  AEACN        label = "Action Taken with Study Treatment"      length = $50
  AEREL        label = "Causality"                              length = $50
  AEOUT        label = "Outcome of Adverse Event"               length = $50
  AESCONG      label = "Congenital Anomaly or Birth Defect"     length = $2
  AESDISAB     label = "Persist or Significant Disability"      length = $2
  AESDTH       label = "Results in Death"                       length = $2
  AESHOSP      label = "Requires or Prolongs Hospitalization"   length = $2
  AESMIE       label = "Other Medically Important Serious Event" length = $2
  AETOXGR      label = "Standard Toxicity Grade"                length = $2
  AESTDTC      label = "Start Date/Time of Adverse Event"       length = $20
  AEENDTC      label = "End Date/Time of Adverse Event"         length = $20
  AEENRTPT     label = "End Relative to Reference Time Point"   length = $40
  AEENTPT      label = "End Reference Time Point"               length = $40
      ;
    set FINAL;
      ;
run;
```

The final AE data set is saved in the SDTM library, as shown in Figure 3.12.

Figure 3.12: SDTM.AE

	STUDYID	DOMAIN	USUBJID	AESEQ	AETERM	AELLT
1	ABC-001	DM	ABC-001-001-001	1	HEADACHE	HEADACHE
2	ABC-001	DM	ABC-001-001-002	1	SORE NECK	NECK PAIN
3	ABC-001	DM	ABC-001-001-003	1	SORE NECK	NECK PAIN
4	ABC-001	DM	ABC-001-001-004	1	SORE NECK	NECK PAIN
5	ABC-001	DM	ABC-001-001-005	1	SORE NECK	NECK PAIN
6	ABC-001	DM	ABC-001-001-006	1	SORE NECK	NECK PAIN
7	ABC-001	DM	ABC-001-001-007	1	SORE NECK	NECK PAIN
8	ABC-001	DM	ABC-001-001-008	1	SORE NECK	NECK PAIN
9	ABC-001	DM	ABC-001-001-009	1	SORE NECK	NECK PAIN
10	ABC-001	DM	ABC-001-001-010	1	HEADACHE	HEADACHE

AELLTCD	AEDECOD	AEPTCD	AEHLT	AEHLTCD	AEHLGT
10019211 HEADACHES		10019211	HEADACHES NEC	10019233	HEADACHES
10028836 NECK PAIN		10028836	MUSCULOSKELETAL AND CONNECTIVE TISSUE DISORDERS	10068757	MUSCULOSKELETAL AND CONNECTIVE TISSUE DISORDERS NEC
10028836 NECK PAIN		10028836	MUSCULOSKELETAL AND CONNECTIVE TISSUE DISORDERS	10068757	MUSCULOSKELETAL AND CONNECTIVE TISSUE DISORDERS NEC
10028836 NECK PAIN		10028836	MUSCULOSKELETAL AND CONNECTIVE TISSUE DISORDERS	10068757	MUSCULOSKELETAL AND CONNECTIVE TISSUE DISORDERS NEC
10028836 NECK PAIN		10028836	MUSCULOSKELETAL AND CONNECTIVE TISSUE DISORDERS	10068757	MUSCULOSKELETAL AND CONNECTIVE TISSUE DISORDERS NEC
10028836 NECK PAIN		10028836	MUSCULOSKELETAL AND CONNECTIVE TISSUE DISORDERS	10068757	MUSCULOSKELETAL AND CONNECTIVE TISSUE DISORDERS NEC
10028836	NECK PAIN	10028836	MUSCULOSKELETAL AND CONNECTIVE TISSUE DISORDERS	10068757	MUSCULOSKELETAL AND CONNECTIVE TISSUE DISORDERS NEC
10028836 NECK PAIN		10028836	MUSCULOSKELETAL AND CONNECTIVE TISSUE DISORDERS	10068757	MUSCULOSKELETAL AND CONNECTIVE TISSUE DISORDERS NEC
10028836 NECK PAIN		10028836	MUSCULOSKELETAL AND CONNECTIVE TISSUE DISORDERS	10068757	MUSCULOSKELETAL AND CONNECTIVE TISSUE DISORDERS NEC
10019211 HEADACHES		10019211	HEADACHES NEC	10019233	HEADACHES

AEHLGTCD	AEBODSYS	AESER	AEACN	AEREL
10019231	NERVOUS SYSTEM DISOR	N	DRUG NOT CHANGED	N
10028393	MUSCULOSKELETAL AND	N	DRUG NOT CHANGED	
10028393	MUSCULOSKELETAL AND	N	DRUG NOT CHANGED	N
10028393	MUSCULOSKELETAL AND	N	DRUG NOT CHANGED	
10028393	MUSCULOSKELETAL AND	N	DRUG NOT CHANGED	N
10028393	MUSCULOSKELETAL AND	N	DRUG NOT CHANGED	N
10028393	MUSCULOSKELETAL AND	N	DRUG NOT CHANGED	
10028393	MUSCULOSKELETAL AND	N	DRUG NOT CHANGED	N
10028393	MUSCULOSKELETAL AND	N	DRUG NOT CHANGED	
10019231	NERVOUS SYSTEM DISOR	N	DRUG NOT CHANGED	Y

AEOUT	AESCONG	AESDISAB	AESDTH	AESTDTC	AEENDTC
RECOVERD OR RESOLVED	N	N	N	2021-01-05	2021-01-05
RECOVERD OR RESOLVED	N	N	N	2021-01-07	2021-01-07
RECOVERD OR RESOLVED	N	N	N	2021-01-07	2021-01-07
RECOVERD OR RESOLVED	N	N	N	2021-01-07	2021-01-07
RECOVERD OR RESOLVED	N	N	N	2021-01-07	2021-01-07
RECOVERD OR RESOLVED	N	N	N	2021-01-07	2021-01-07
RECOVERD OR RESOLVED	N	N	N	2021-01-07	2021-01-07
RECOVERD OR RESOLVED	N	N	N	2021-01-07	2021-01-07
RECOVERD OR RESOLVED	N	N	N	2021-01-09	2021-01-09
RECOVERD OR RESOLVED	N	N	N	2021-01-09	2021-01-09

AESHOSP	AESMIE	AETOXGR	AEENRTPT	AEENTPT
N	N	1		
N	N	1		
N	N	2		
N	N	2		
N	N	2		
N	N	2		
N	N	2		
N	N	1		
N	N	1		
N	N	1		

3.6.4 Exposure (EX)

The Exposure (EX) domain describes the details of subjects' exposure to the protocol-specified study treatment. The purpose of the EX domain is to facilitate simple and easy statistical analysis for exposure to treatment.

Note: The Exposure as Collected (EC) domain is introduced to provide comprehensive exposure information and to be part of the trail for Exposure (EX) domain. If the study is blinded, the EC domain is used rather than EX domain. In this study, we provide an example of how to generate EX domain because the study design is open label.

Structure of EX

The structure of EX consists of Identifier, Timing, Topic, and Qualifier variables, as follows.

Identifier Variables:

- STUDYID: Required variable in character format, the unique study identifier. Each study has a unique study ID. In the example study described in this book, we populate it with "ABC-001".
- DOMAIN: Required variable in character format, compliant with SDTM Implementation Guide. *DOMAIN* sets to "EX".
- USUBJID: Required variable in character format used to uniquely identify a subject across all studies. The *USUBJID* variable is derived using: *STUDYID- SITED-SUBJID*, such as "ABC-001-001-001".
- EXSEQ: Required variable in numeric format, the sequence number given to ensure uniqueness of subject records within a domain.
- EXREFID: Permissible variable in character format, the internal or external identifier.

Timing Variables:

- EXSTDTC: Required variable in character format, the start date and time of treatment.
- EXENDTC: Required variable in character format, the end date and time of treatment.

Topic Variables:

- EXTRT: Required variable in character format, the name of treatment.

Qualifier Variables:

- EXCAT: Permissible variable in character format, the description of the *EXTRT* values.
- EXDOSTXT: Expected variable in character format, the amount of *EXTRT* when *EXTRT* is character.
- EXDOSE: Expected variable in character format, the amount of *EXTRT* when *EXTRT* is numeric.
- EXDOSU: Expected variable in character format, the units for *EXDOSE, EXDOSETXT* in compliance with protocol-specified value.
- EXDOSFRM: Expected variable in character format, the dose form for *EXTRT*.
- EXDOSFRQ: Permissible variable in character format, the number of repeated administrations of *EXDOSE* within a specific time period.
- EXROUTE: Permissible variable in character format, the route of administration for the intervention.

EX eCRF Reference

Exposure information is collected during the course of treatment. In practice, exposure information is recorded on one or more eCRF pages.

Figure 3.13: SDTM Annotated Exposure (EX)

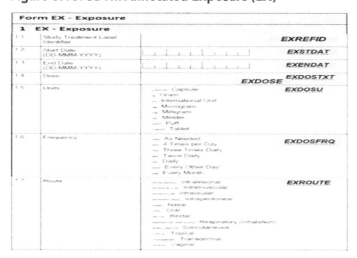

EX Specification

Table 3.5: EX Specification

Variable	Label	Type	Length	Control Terminology	Origin	Notes
STUDYID	Study Identifier	char	20		Protocol: Set to "ABC-001"	This information that programmer can find in multiple raw data sets, such as RAW.DM
DOMAIN	Domain Abbreviation	char	2	DOMAIN	Assigned: Set to "EX"	
USUBJID	Unique Subject Identifier	char	40		Derived: Concatenate among three variables below: STUDYID-SITED-SUBJID	
EXSEQ	Sequence number	num	8		Derived: Sequence number given to ensure uniqueness of subject records	Sort by USUBJID, EXSTDTC, EXENDTC, EXTRT, EXDOSE, starts with value 1 for the first row of each subject, then increment 1 at each successive record for each subject.
EXREFID	Reference ID	char	100			
EXTRT	Name of Treatment	char	100		CRF	Reference raw EX.EXTRT

(Continued)

Table 3.5: (*Continued***)**

Variable	Label	Type	Length	Control Terminology	Origin	Notes
EXCAT	Category of Treatment	char	40		Assigned	Set to "DOSING"
EXDOSE	Dose	num	8		CRF.	Raw EX.EXDOSE
EXDOSU	Dose Units	char	40		CRF	Raw EX.EXDOSU
EXDOSTXT	Dose Description	char	40		CRF	Raw EX.EXDOSTXT
EXDOSFRM	Dose Form	char	20		Assigned	Set to "TABLET"
EXDOSFRQ	Dosing Frequency per Interval	char	20		CRF	Raw EX.EXDOSFRQ
EXROUTE	Route of Administration	char	40		CRF	RAW EX.EXROUTE
EXSTDTC	Start Date/ Time of Treatment	char	40		CRF	Raw EX.EXSTDAT
EXENDTC	End Date/Time of Treatment	char	40		CRF	Raw EX.EXENDAT

EX Programming

In order to create SDTM.EX, we first run a PROC CONTENTS on the raw data. Please see the code below.

Note: For every domain, the corresponding SAS name is domain.sas. For example, EX is created by ex.sas, so all code in this section belongs to ex.sas.

```
/*Begin writing SAS program ex.sas*/
/*show structure of the raw EX dataset*/
proc contents data=RAW.EX;
run;
```

First, as seen in Figure 3.14, we use PROC CONTENTS to see that there are 14 variables in the raw EX data set. Further, we can see the Type, Length (Len), Format, and Informat of each variable. These variable attributes are assigned by SAS based on the raw data format. The Label is assigned in SAS based on the first row of the raw EX data set in the Excel sheet.

Note: The purpose of running PROC CONTENTS is to get a clear idea of what the raw metadata looks like and compare how similar it is to the SDTM domain that we will create from it.

Figure 3.14: Alphabetic List of Variables and Attributes

Alphabetic List of Variables and Attributes

#	Variable	Type	Len	Format	Informat	Label
2	COUNTRY	Char	13	$13	$13	COUNTRY
7	ESEQ	Num	8	BEST		ESEQ
5	EVENT	Char	24	$24	$24	EVENT
6	EVENT_DT	Num	8	MMDDYY10		EVENT_DT
14	EXDOSE	Num	8	BEST		EXDOSE
17	EXDOSFRQ	Char	6	$6	$6	EXDOSFRQ
16	EXDOSTXT	Char	1	$1	$1	EXDOSTXT
15	EXDOSU	Char	2	$2	$2	EXDOSU
13	EXENDAT	Num	8	DATETIME16		EXENDAT
8	EXREFID	Num	8	BEST		EXREFID
18	EXROUTE	Char	4	$4	$4	EXROUTE
12	EXSTDAT	Num	8	DATETIME16		EXSTDAT
9	FORM	Char	26	$26	$26	FORM
10	FSEQ	Num	8	BEST		FSEQ
11	IGSEQ	Num	8	BEST		IGSEQ
3	SITEID	Char	3	$3	$3	SITEID
1	STUDYID	Char	7	$7	$7	STUDYID
4	SUBJID	Char	3	$3	$3	SUBJID

```
/*Create the 1st set of EX variables using existing variables from RAW.EX*/
data EX1;
  /*Specify length for standard variables*/
  length EXSTDTC EXENDTC EXREFID $20. EXTRT $100;
  /*Rename EXDOSU, EXDOSE, EXDOSFRQ, EXROUTE, EXREFID*/
  set RAW.EX(rename=(EXDOSU=EXDOSU_ EXDOSE=EXDOSE_ EXDOSFRQ=EXDOSFRQ_ EX-
ROUTE=EXROUTE_  EXREFID=EXREFID_));

  /*Define DOMAIN, USUBJID, EXCAT, EXDOSFRM */
  /*Derive EXTRT, EXDOSU, EXDOSE, EXDOSFRQ, EXROUTE, EXREFID, EXDTS, EXTMS */
  DOMAIN="EX";    ❶
  USUBJID=STRIP(STUDYID)||"-"||STRIP(SITEID)||"-"||STRIP(SUBJID);
  if index(upcase(FORM),"DRUG A")>0 then EXTRT="DRUG A";
  EXCAT="DOSING";
  EXDOSU=upcase(strip(EXDOSU_));
  EXDOSE=EXDOSE_;
  EXDOSFRM="TABLET";
  EXDOSFRQ=upcase(strip(EXDOSFRQ_));
  EXROUTE=upcase(strip(EXROUTE_));
  EXROUTE=upcase(strip(EXROUTE_));
  EXDTS=datepart(EXSTDAT);
  EXTMS=timepart(EXSTDAT);

  /*Format EXDTS_DT, EXDTS_TM, EXDTE_DT, EXDTE_TM */
  /*Derive EXDTE, EXTME, EXSTDTC, EXENDTC */
  EXDTS_DT=put(EXDTS,yymmdd10.);
  EXDTS_TM=put(EXTMS,time8.);
```

```
   EXDTE=datepart(EXENDAT);        ❷
   EXTME=timepart(EXENDAT);
   EXDTE_DT=put(EXDTE,yymmdd10.);
   EXDTE_TM=put(EXTME,time8.);

   EXSTDTC=strip(EXDTS_DT)||"T"||strip(EXDTS_TM);
   EXENDTC=strip(EXDTE_DT)||"T"||strip(EXDTE_TM);
run;
```

❶ Create variables: DOMAIN, USUBJID, EXCAT, EXTRT, EXDOSE, EXDOSU, EXDOSFRM, EXDOSRQ, EXROUTE, EXSTDTC, and EXENDTC.

❷ Be diligent when you create variables with date and time, such as EXSTDTC and EXENDTC, which are derived from the variables EXSTDAT and EXENDAT (numeric) in the RAW.EX data set. We use put datepart() and timepart() to split the date and time part, then use put() with yymmdd10. and put() with time8. to switch numeric value to character format. Finally, we use concatenation || to combine date and time parts with "T".

```
/*Sort data set EX1 by USUBJID, EXSTDTC, EXENDTC, EXTRT, EXDOSE */
proc sort data=EX1 out=EX2;
    by USUBJID EXSTDTC EXENDTC EXTRT EXDOSE ;
run;

data Final;
  set EX2;
/*Derive EXSEQ*/

  length EXSEQ 8.;     ❸
  by USUBJID EXSTDTC EXENDTC EXTRT EXDOSE ;
  if FIRST.USUBJID then EXSEQ=0;
  EXSEQ+1;
  output;
  format _all_;
  informat _all_;
run;
```

❸ Create the EXSEQ variable, which is used to uniquely record the sequence number of a subject within a domain. To create EXSEQ, please refer to the detailed explanation for the DSSEQ variable under DS domain in Section 3.6.2.

```
libname SDTM "…/directory";
data SDTM.EX(label="Exposure");

/*Assign variable attributes such as label and length to conform with SDTM.EX
Specification (these will also be the same attributes as the SDTM IG).*/
attrib

        STUDYID        label = "Study Identifier"              length = $20
        DOMAIN         label = "Domain Abbreviation"           length = $2
        USUBJID        label = "Unique Subject Identifier"     length = $40
        EXSEQ          label = "Sequence Number"               length = 8
        EXREFID        label = "Reference ID"                  length = $20
        EXTRT          label = "Name of Treatment"             length = $100
```

```
        EXCAT           label = "Category of Treatment"            length = $40
        EXDOSE          label = "Dose"                             length = 8
        EXDOSU          label = "Dose Units"                       length = $40
        EXDOSFRM        label = "Dose Form"                        length = $20
        EXDOSFRQ        label = "Dosing Frequency per Interval"     length = $20
        EXROUTE         label = "Route of Administration"          length = $40
        EXSTDTC         label = "Start Date/Time of Treatment"     length = $40
        EXENDTC         label = "End Date/Time of Treatment"       length = $40
        ;
    set Final;

    keep STUDYID DOMAIN USUBJID EXSEQ EXREFID EXTRT EXCAT EXDOSE EXDOSU
EXDOSFRM EXDOSFRQ EXROUTE EXSTDTC EXENDTC;
run;
```

The final EX data set is saved in the SDTM library, as shown in Figure 3.15.

Figure 3.15: SDTM.EX

3.6.5 Concomitant Medications (CM)

An interventions domain contains concomitant and prior medications used by the subject. CM domain is one record per medication intervention taken.

Structure of CM

The Concomitants Medications domain contains data about non-study drugs taken during the course of study treatment. The structure of CM consists of Identifier, Timing, Topic, and Qualifier variables, as follows.

Identifier Variables:

- STUDYID: Required variable in character format, the unique study identifier. Each study has a unique study id. In the example study described in this book, we populate it with "ABC-001".
- DOMAIN: Required variable in character format, compliant with SDTM Implementation Guide. DOMAIN sets to "CM".
- USUBJID: Required variable in character format used to uniquely identify a subject across all studies. The *USUBJID* variable is derived using: *STUDYID- SITED-SUBJID*, such as "ABC-001-001-001".
- CMSEQ: Required variable in numeric format, the sequence number given to ensure uniqueness of subject records within a domain.

Timing Variables:

- CMSTDTC: Required variable in character format the start date and time of Medication, representing in ISO 8601 character format.
- CMENDTC: Required variable in character format the end date and time of Medication, representing in ISO 8601 character format.
- CMENRTPT: Permissible variable in character format, the end relative to reference time point.

Topic Variables:

- CMTRT: Required variable in character format, the reported name of drug, medication, or therapy.

Qualifier Variables:

- CMDECOD: Required variable in character format, the dictionary-derived text description of reported name (CMTRT), equivalent to the genetic drug name in WHO Drug.
- CMCAT: Permissible variable in character format, the description of a category of medications and treatments. For example: "PRIOR", "CONCOMITANT", "GENERAL CONMED".

- CMINDC: Permissible variable in character format, the description of why a medication was taken or administrated.
- CMROUTE: Permissible variable in character format, the route of administration for the intervention. For example: "ORAL", "INTRAVENOUS".

CM eCRF Reference

Prior and Concomitant Medications information is collected in each stage including "Screening", "Treatment", and "Follow-Up. CM records information related to concomitant and prior medications used by subject to produce a therapeutic effect during the protocol- specified collection episode, including transfusion of bloods or blood products.

Figure 3.16: SDTM Annotated Prior and Concomitant Medications (CM)

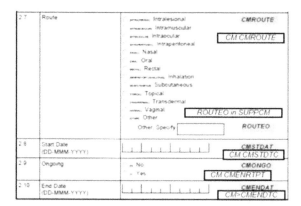

CM Specification

Table 3.6: CM Specification

Variable	Label	Type	Length	Control Terminology	Origin	Notes
STUDYID	Study Identifier	char	20		Protocol: Set to "ABC-001"	This information that programmer can find in multiple raw data sets, such as RAW.IC; RAW.DM; RAW.AE, etc.

(Continued)

Table 3.6: (*Continued*)

Variable	Label	Type	Length	Control Terminology	Origin	Notes
DOMAIN	Domain Abbreviation	char	2	DOMAIN	Assigned: Set to "CM"	
USUBJID	Unique Subject Identifier	char	40		Derived: Concatenate among three variables below: STUDYID-SITED-SUBJID	
CMSEQ	Sequence Number	num	8		Derived: Sequence number given to ensure uniqueness of subject records	Sort by USUBJID, CMCAT, CMSTDTC, CMENDTC, CMTRT, starts with value 1 for the first row of each subject, then increment 1 at each successive record for each subject.
CMTRT	Reported Name of Drug, Medication or Therapy	char	100		CRF.	Raw data set CM.CMTRT, Uppercase the value
CMDECOD	Standardized Medication Name	char	100	WHODrug	Assigned: Assigned per MedDRA dictionary.	Raw data set CM.CMDECOD
CMDOSU	Dose Units	char	40		CRF	Raw data set CM.CMDOSU
CMDOSFRM	Dose Form	char	40		CRF	Raw data set CM.CMDOSFRM

(*Continued*)

Table 3.6: (*Continued***)**

Variable	Label	Type	Length	Control Terminology	Origin	Notes
CMDOSFRQ	Dose Frequency	char	40		CRF	Raw data set CM.CMDOSFRQ
CMCAT	Category for Medication	char	40	CMCAT	CRF	Set to "General"
CMINDC	Indication	char	200		CRF.	Raw data set CM.CMINDC. Use uppercase for the value
CMROUTE	Route of Administration	char	40		CRF	Raw data set CM.CMROUTE
CMSTDTC	Start Date/ Time of Medication	char	20		CRF	Raw data set CM.CMSTDAT
CMENDTC	End Date/ Time of Medication	char	20		CRF	Raw data set CM.CMENDAT
CMENRTPT	End Relative to Reference Time Point	char	20		CRF	Set to "ONGOING" if raw data set CM.CMONGO is not missing

CM Programming

In order to create SDTM.CM, we first run a PROC CONTENTS on the raw data. Please see the code below.

Note: For every domain, the corresponding SAS name is domain.sas. For example, CM is created by cm.sas, so all code in this section belongs to cm.sas.

```
/* Begin writing SAS program cm.sas*/
/*show structure of the raw CM dataset*/
proc contents data=RAW.CM;
run;
```

First, as seen in Figure 3.17, we use PROC CONTENTS to see that there are 25 variables in the raw CM data set. Further, we can see the Type, Length (Len), Format, and Informat of each variable. These variable attributes are assigned by SAS based on the raw data format. The Label is assigned in SAS based on the first row of the raw CM data set in the Excel sheet.

Note: The purpose of running PROC CONTENTS is to get a clear idea of what the raw metadata looks like and compare how similar it is to the SDTM domain that we will create from it.

Figure 3.17: Alphabetical List of Variables and Attributes

Alphabetic List of Variables and Attributes

#	Variable	Type	Len	Format	Informat	Label
17	CMDECOD	Char	9	$9	$9	CMDECOD
18	CMDOSFRM	Char	7	$7	$7	CMDOSFRM
19	CMDOSFRQ	Char	6	$6	$6	CMDOSFRQ
20	CMDOSU	Char	17	$17	$17	CMDOSU
14	CMENDAT	Num	8	MMDDYY10		CMENDAT
25	CMINDC	Char	13	$13	$13	CMINDC
15	CMONGO	Char	1	$1	$1	CMONGO
16	CMROUTE	Char	6	$6	$6	CMROUTE
13	CMSTDAT	Num	8	MMDDYY10		CMSTDAT
12	CMTRT	Char	12	$12	$12	CMTRT
11	CMYN	Char	1	$1	$1	CMYN
2	COUNTRY	Char	13	$13	$13	COUNTRY
22	DOSFRMO	Char	1	$1	$1	DOSFRMO
23	DOSFRMQO	Char	4	$4	$4	DOSFRMQO
21	DOSUO	Char	1	$1	$1	DOSUO
6	ESEQ	Num	8	BEST		ESEQ
5	EVENT	Char	4	$4	$4	EVENT
7	FORM	Char	33	$33	$33	FORM
8	FORMO	Char	2	$2	$2	FORMO
9	FSEQ	Num	8	BEST		FSEQ
10	XGSEQ	Num	8	BEST		XGSEQ
24	ROUTEO	Char	1	$1	$1	ROUTEO
3	SITEID	Char	3	$3	$3	SITEID
1	STUDYID	Char	7	$7	$7	STUDYID
4	SUBJID	Char	3	$3	$3	SUBJID

```
/*Create the 1st set of CM variables using existing variables from RAW.CM*/
data CM1;
/*Specify length for standard variables*/
  length STUDYID CMSTDTC CMENDTC CMENRTPT $20. DOMAIN $2. USUBJID CMCAT
CMROUTE $40. CMTRT CMINDC $200.;
  /*Rename STUDYID, CMTRT, CMINDC, CMDOSU, CMROUTE, CMDOSFRM, CMDOSFRQ,
CMDECOD */
  set RAW.CM(where=(CMYN="Y") rename=(STUDYID=STUDYID_ CMTRT=CMTRT_
CMINDC=CMINDC_ CMDOSU=CMDOSU_ CMROUTE=CMROUTE_ CMDOSFRM=CMDOSFRM_
CMDOSFRQ=CMDOSFRQ_ CMDECOD=CMDECOD_));   ❶
  /*Define DOMAIN, CMCAT*/
  DOMAIN="CM";
  /*Derive STUDYID, USUBJID, CMROUTE, CMDOSU, CMDOSFRM, CMDOSFRQ */
  STUDYID=strip(STUDYID_);
  USUBJID=STRIP(STUDYID)||"-"||STRIP(SITEID)||"-"||STRIP(SUBJID);
  CMTRT=strip(CMTRT_);
  CMCAT="GENERAL";
  CMROUTE=strip(upcase(CMROUTE_));
  CMDOSU=strip(upcase(CMDOSU_));
```

```
    CMDOSFRM=strip(upcase(CMDOSFRM_));
    CMDOSFRQ=strip(upcase(CMDOSFRQ_));
    /*Format CMSTDTC, CMENDTC */
    CMSTDTC=put(CMSTDAT,yymmdd10.);        ❷
    CMENDTC=put(CMENDAT,yymmdd10.);
    /*Derive CMTRT, CMINDC, CMDECOD, CMENRTPT */
    CMTRT=strip(upcase(CMTRT_));
    CMINDC=strip(upcase(CMINDC_));
    CMDECOD=strip(upcase(CMDECOD_));
    if CMONGO^="" then CMENRTPT="ONGOING";
run;
```

❶ The WHERE statement is used to filter the condition when *CMYN*= "Y" (Any Medications Were Taken?). Next, create variables: *DOMAIN, USUBJID, CMTRT, CMCAT, CMROUTE, CMDOSU, CMDOSFRM, CMDOSFRQ, CMSTDTC, CMENDTC, CMINDC, CMDECOD, CMENRTPT.*

❷ Be diligent when you create variables with date and time, such as *CMSTDTC* and *CMENDTC*, which are derived from the variables CMSTDAT and CMENDAT (numeric) in the RAW.CM data set. We use put() function with yymmdd10. Format to switch numeric values to character in yyyy-mm-dd format.

```
/*Sort data set CM1 by USUBJID, CMCAT, CMSTDTC, CMENDTC, CMTRT */
proc sort data=CM1 out=CM2;
    by USUBJID CMCAT CMSTDTC CMENDTC CMTRT;
run;

data Final;
    set CM2;
/*Derive CMSEQ*/
    length CMSEQ 8.;
    by USUBJID CMCAT CMSTDTC CMENDTC CMTRT;
    if FIRST.USUBJID then CMSEQ=0;        ❸
      CMSEQ+1;
      output;
run;
```

❸ Create the *CMSEQ* variable, which is used to uniquely record the sequence number of a subject within a domain. To create *CMSEQ*, please refer to the detailed explanation for *DSSEQ* variable under DS domain in Section 3.6.2.

```
libname SDTM ".../directory";
data SDTM.CM(label="Concomitant Medications");
/*Assign variable attributes such as label and length to conform with SDTM.CM
Specification (these will also be the same attributes as the SDTM IG).*/
Attrib
  STUDYID    label = "Study Identifier"                            length = $20
  DOMAIN     label = "Domain Abbreviation"                         length = $2
  USUBJID    label = "Unique Subject Identifier"                   length = $40
  SUBJID     label = "Subject Identifier for the Study"            length = $20
  CMSEQ      label = "Sequence Number"                             length = 8
  CMTRT      label = "Reported name of drug, Medication or Therapy" length = $200
```

```
    CMDECOD      label = "Standardized Medication Name"            length = $100
    CMCAT        label = "Category for Medication"                 length = $40
    CMINDC       label = "Indication"                              length = $200
    CMDOSU       label = "Dose Units"                              length = $40
    CMDOSFRM     label = "Dose Form"                               length = $40
    CMDOSFRQ     label = "Dose Frequency"                          length = $40
    CMROUTE      label = "Route of Administration"                 length = $40
    CMSTDTC      label = "Start Date/Time of Medication"           length = $20
    CMENDTC      label = "End Date/Time of Medication"             length = $20
    CMENRTPT     label = "End Relative to Reference Time Point"    length = $20


       ;
    set Final;
      keep STUDYID DOMAIN USUBJID SUBJID CMSEQ CMTRT CMDECOD CMCAT CMDOSU
CMDOSFRM CMDOSFRQ CMINDC CMROUTE CMSTDTC CMENDTC CMENRTPT
       ;
run;
```

The final CM data set is saved in the SDTM library, as shown in Figure 3.18.

Figure 3.18: SDTM.CM

	STUDYID	DOMAIN	USUBJID	SUBJID	CMSEQ	CMTRT
1	ABC-001	CM	ABC-001-001-001	001	1	PRUNE JUICE
2	ABC-001	CM	ABC-001-001-002	002	1	PRUNE JUICE
3	ABC-001	CM	ABC-001-001-006	006	1	PRUNE JUICE
4	ABC-001	CM	ABC-001-001-009	009	1	ASPIRIN

CMTRT	CMDECOD	CMCAT	CMINDC	CMDOSU
PRUNE JUICE	PRUNELLA	GENERAL	CONSTIPATION	LITER PER MINUTE
PRUNE JUICE	PRUNELLA	GENERAL	CONSTIPATION	LITER PER MINUTE
PRUNE JUICE	PRUNELLA	GENERAL	CONSTIPATION	LITER PER MINUTE
ASPIRIN	PRUNELLA	GENERAL	FEVER	MILLIGRAM

CMDOSFRM	CMDOSFRQ	CMROUTE	CMSTDTC	CMENDTC	CMENRTPT
LIQUID	DAILY	ORAL	2021-01-06	2021-01-06	
LIQUID	DAILY	ORAL	2021-01-06	2021-01-06	
LIQUID	DAILY	ORAL	2021-01-07	2021-01-07	
LIQUID	OTHER	ORAL	2021-01-09	2021-01-09	

3.6.6 Electrocardiogram Test Results (EG)

Electrocardiogram is the most commonly recorded physiological signal. It is a vital component of all diagnostic tests involving cardiac health and includes parameters derived from the ECG waveform such as Heart Rate (HR), PR Interval, QRS Interval, QT Interval, and RR Interval.

Structure of EG

The structure of EG consists of Identifier, Timing, Topic, and Qualifier variables, as follows.

Identifier Variables:

- STUDYID: Required variable in character format, the unique study identifier. Each study has a unique study id. In the example study described in this book, we populate it with "ABC-001".
- DOMAIN: Required variable in character format, compliant with SDTM Implementation Guide. *DOMAIN* sets to "EG".
- USUBJID: Required variable in character format used to uniquely identify a subject across all studies. The *USUBJID* variable is derived using: *STUDYID- SITED-SUBJID*, such as "ABC-001-001-001".
- EGSEQ: Required variable in numeric format, the sequence number given to ensure uniqueness of subject records within a domain.
- EGREFID: Permissible variable in character format, the ECG Reference ID.

Timing Variables:

- EGDTC: Required variable in character format, the Date and Time of ECG.
- VISIT: Permissible variable in character format, the Visit Name.
- VISITNUM: Expected variable in character format, the Visit Number.

Topic Variables:

- EGTESTCD: Required variable in character format, the ECG Test or Examination Short Name.

Qualifier Variables:

- EGTEST: Required variable in character format, the ECG Test or Examination Name.
- EGCAT: Permissible variable in character format, the Category for ECG.
- EGPOS: Permissible variable in character format, the ECG Position of Subject.
- EGORRES: Expected variable in character format, the result of Finding in Original Units.
- EGORRESU: Expected variable in character format, the Original Units.
- EGMETHOD: Permissible variable in character format, the Method of ECG Test.
- EGSTRESC: Expected variable in character format, the Character Result of Finding in Std Format.
- EGBLFL: Expected variable in character format, the Baseline Flag.

EG eCRF Reference

Electrocardiogram information is collected in stages including "Screening" and "Treatment". The electrocardiogram data collected includes information as follows: position of the subject, method of measurement, type of measurement, and overall interpretation.

In terms of the CDISC eCRF, there are two types of ECG, including Central Reading (Central) and Local Reading (Local) ECG, and each are captured from a different vendor.

Note: For the Central ECG, the results are captured directly by a central vendor with an electronic device. The results are not typically recorded directly into the eCRF. In our simulated study, the Central ECG is collected on the eCRF. For the Local ECG, results are captured by a local vendor, and the results can be recorded directly in the eCRF.

If performing multiple ECG measurements on the same day, the planned time point is collected using variable EGTPT. In this study design, there is one ECG measurement for each day for three consecutive treatment days to be consistent with the eCRF design as shown in Figure 3.19.

Figure 3.19: SDTM Annotated Electrocardiogram (EG)

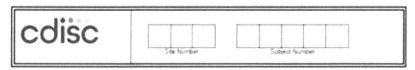

Form EG - Central Reading		
1 EG - Central Reading		
1 1	Was an ECG performed?	⦿ No EG EGPERF **EGPERF** ⦿ Yes NOT SUBMITTED
1 2	ECG Reference Identifier/ Accession Number	EG EGREFID **EGREFID**
1 3	Method	⦿ 12 LEAD STANDARD **EGMETHOD** ⦿ HOLTER CONTINUOUS ECG RECORDING EG EGMETHOD
1 4	Position	⦿ Sitting **EGPOS** ⦿ Standing ⦿ Supine EG EGPOS ⦿ Semi-Recumbent ⦿ Semi-Fowler's
1 5	Date (DD-MMM-YYYY)	|__|__|__|__|__|__|__| **EGDAT** EG EGDTC

EG Specification

Table 3.7: EG Specification

Variable	Label	Type	Length	Control Terminology	Origin	Notes
STUDYID	Study Identifier	char	20		Protocol: Set to "ABC-001"	This information that programmer can find in multiple raw data sets, such as RAW.IC; RAW.DM; RAW.AE, etc.
DOMAIN	Domain Abbreviation	char	2	DOMAIN	Assigned: Set to "EG"	

(Continued)

Table 3.7: (*Continued*)

Variable	Label	Type	Length	Control Terminology	Origin	Notes
USUBJID	Unique Subject Identifier	char	40		Derived: Concatenate among three variables below: STUDYID-SITED-SUBJID	
EGSEQ	Sequence Number	num	8		Derived: Sorted by USUBJID, EGCAT, EGTESTCD, EGDTC, VISITNUM, EGTPTNUM, EGSPID	
EGREFID	ECG Reference ID	char	20			Raw EG.ESEQ
EGTESTCD	ECG Test or Examination Short Name	char	8		Assigned: Short Name of the measurement test, or examination described in EGTEST.	HR, PR, QRS, QT, QTCA
EGTEST	ECG Test or Examination Short Name	char	20		Verbatim name of the test or examination used to obtain the measurement or finding.	

(*Continued*)

Table 3.7: (*Continued*)

Variable	Label	Type	Length	Control Terminology	Origin	Notes
EGCAT	Category of ECG	char	40		Assigned	Set "MEASUREMENT" when EGTESTCD is equal to "HR"; set "INTERVAL" when EGTESTCD is equal to "PR", "QRS", "QT", "QTca". Set to "NULL" under other conditions.
EGPOS	ECG Position of Subject	char	20			Position of the Subject during a measurement or examination. Set to "NULL" when EGTEST is equal to "RR" and "QTCF". Set to "SUPINE" for all other tests.
EGORRES	Result or Finding in Original Units	Char	100		Assigned	Result of ECG measurement or finding as originally received or collected.
EGORRESU	Original Units	char	40		Assigned	Original Units in which the data was collected
EGSTRESC	Character Result/Finding in Standard Format	char	100			Contains the result value for all findings, copied, or derived from EGORRES in a standard format or unit.
EGMETHOD	Method of ECG Test	char	40			Raw EG.EGMETHOD

(*Continued*)

Table 3.7: (*Continued*)

Variable	Label	Type	Length	Control Terminology	Origin	Notes
EGBLFL	Baseline Flag	char	1		Derived	Indicator used to identify a baseline value. Set to "Y" for the last nonmissing result for each ECG test that occurred before first study dose, EX.EXSTDTC as reference
VISIT	Visit Name	char	20			Map from ECG. EVENT
VISITNUM	Visit Number	num	8		Assigned	Map VISITNUM from SDTM.TV
EGDTC	Date/Time of ECG	char	20			Raw EG.EGDTC

EG Programming

In order to create SDTM.EG, we first run a PROC CONTENTS on the raw data. Please see the code below.

Note: For every domain, the corresponding SAS name is domain.sas. For example, EG is created by eg.sas, so all code in this section belongs to eg.sas.

```
/*Begin writing SAS program eg.sas*/
/*show structure of the raw EG dataset*/
proc contents data=RAW.EG;
run;
```

First, as seen in Figure 3.20, we use PROC CONTENTS to see that there are 25 variables in the raw EG data set. Further, we can see the Type, Length (Len), Format, and Informat of each variable.

Figure 3.20: Alphabetic List of Variables and Attributes

Alphabetic List of Variables and Attributes

#	Variable	Type	Len	Format	Informat	Label
2	COUNTRY	Char	13	$13	$13	COUNTRY
12	EGDAT	Num	8	DATETIME16		EGDAT
14	EGHRMN_EGORRES	Num	8	BEST		EGHRMN_EGORRES
15	EGHRMN_EGORRESU	Char	9	$9	$9	EGHRMN_EGORRESU
13	EGMETHOD	Char	17	$17	$17	EGMETHOD
25	EGORRES	Char	1	$1	$1	EGORRES
7	ESEQ	Num	8	BEST		ESEQ
5	EVENT	Char	17	$17	$17	EVENT
6	EVENTDT	Char	1	$1	$1	EVENTDT
8	FORM	Char	17	$17	$17	FORM
9	FORMEID	Char	1	$1	$1	FORMEID
10	FSEQ	Num	8	BEST		FSEQ
11	IGSEQ	Num	8	BEST		IGSEQ
24	INTP_EGORRES	Char	7	$7	$7	INTP_EGORRES
16	PRSB_EGORRES	Num	8	BEST		PRSB_EGORRES
17	PRSB_EGORRESU	Char	4	$4	$4	PRSB_EGORRESU
18	QRSAG_EGORRES	Num	8	BEST		QRSAG_EGORRES
19	QRSAG_EGORRESU	Char	5	$5	$5	QRSAG_EGORRESU
20	QTAG_EGORRES	Num	8	BEST		QTAG_EGORRES
21	QTAG_EGORRESU	Char	5	$5	$5	QTAG_EGORRESU
22	QTCAAG_EGORRES	Num	8	BEST		QTCAAG_EGORRES
23	QTCAAG_EGORRESU	Char	4	$4	$4	QTCAAG_EGORRESU
3	SITEID	Char	3	$3	$3	SITEID
1	STUDYID	Char	7	$7	$7	STUDYID
4	SUBJID	Char	3	$3	$3	SUBJID

These variable attributes are assigned by SAS based on the raw data format. The Label is assigned in SAS based on the first row of the raw EG data set in the Excel sheet.

Note: The purpose of running PROC CONTENTS is to get a clear idea of what the raw metadata looks like and compare how similar it is to the SDTM domain that we will create from it.

We use PROC FORMAT to create a format that converts the character strings on the left side to the corresponding values on the right side. For readers unfamiliar with CDISC standards, using this format helps us map *EGTEST* to a corresponding character variable *EGTESTCD* as follows: *HR* to Heart Rate, *INTP* to Interpretation, *PR* to PR Interval, *QT* to QT Interval, and *QTca* to QTca Interval.

```
/*Assign the character values from "HR", "INTP", "PR", "QRS", "QT", "QTca"
to character values "Heart Rate", "Interpretation", "PR Interval", "QRS
```

```
Interval", "QT Interval" and "QT Interval" for variable EGTESTCD using PROC
FORMAT*/
proc format;
  value $EGTESTCD
    "HR"="Heart Rate"
       "INTP"="Interpretation"
       "PR"="PR Interval"
       "QRS"="QRS Interval"
       "QT"="QT Interval"
       "QTca"="QTca Interval"
       ;
quit;
```

The code below creates variables *HR, PR, QRS, QT, QTca* in order to prepare to transpose the data in the following step. Derive *VISIT* based on *EVENT* from RAW.EG, as shown in Figure 3.21.

```
data EG1;
  /*Specify length for standard variables*/
  length INTP $300. USUBJID $40. HR PR QRS QT QTca $200. VISIT $40.;

  /*Rename STUDYID EGMETHOD*/
  set RAW.EG(rename=(STUDYID=STUDYID_ EGMETHOD=EGMETHOD_));
  /*Define DOMAIN*/
  DOMAIN="EG";
  /*Derive STUDYID, USUBJID, EGREFID */
  STUDYID=strip(STUDYID_);
  USUBJID=STRIP(STUDYID)||"-"||STRIP(SITEID)||"-"||STRIP(SUBJID);

  EGMETHOD=upcase(strip(EGMETHOD_));
  EGREFID=strip(put(FSEQ,best.));

  /*Derive HR, PR, QRS, QT, QTca, INTP*/
  HR=catx("~",Coalescec(put(EGHRMN_EGORRES,best.),"_null_"),EGHRMN_EGORRESU);
❶
  PR=catx("~",Coalescec(put(PRSB_EGORRES,best.),"_null_"),PRSB_EGORRESU);
  QRS=catx("~",Coalescec(put(QRSAG_EGORRES,best.),"_null_"),QRSAG_EGORRESU);
  QT=catx("~",Coalescec(put(QTAG_EGORRES,best.),"_null_"),QTAG_EGORRESU);
  QTca=catx("~",Coalescec(put(QTCAAG_EGORRES,best.),"_null_"),QTCAAG_
EGORRESU);
  INTP=catx("~",INTP_EGORRES,EGORRES);
  /*Derive VISIT*/
  if index(upcase(EVENT),"SCREENING") then VISIT="SCREENING";
  else if index(upcase(EVENT),"FOLLOW UP") then VISIT="FOLLOW-UP";
  else if index(upcase(EVENT),"DAY") then VISIT="DAY
"||strip(substr(upcase(EVENT),length(EVENT)-1));
  else VISIT=strip(upcase(EVENT));

run;
```

❶ The purpose of Coalesces() function is to check each expression in order and return the first nonmissing value, then CATX() is to concatenate two strings with "~" symbol.

```
  /*Sort data set EG1 by STUDYID, DOMAIN, USUBJID, EGREFID, VISIT, EGDAT */
proc sort data=EG1;
    by STUDYID DOMAIN USUBJID EGMETHOD EGREFID VISIT EGDAT ;
run;
  /*Transpose data set EG1 with HR, PR, QRS, QT, QTca, INTP */
proc transpose data=EG1 out=EG2;     ❷
    by STUDYID DOMAIN USUBJID EGMETHOD EGREFID VISIT EGDAT;
    var HR PR QRS QT QTca INTP;
run;
```

❷ Transpose the HR, PR, QRS, QT, QTca, and INTP data sets from horizontal structure to vertical structure.

Figures 3.21 and 3.22 show the output EG1 (before PROC TRANSPOSE) and output EG2 (after PROC TRANSPOSE) to address how PROC TRANSPOSE changes the data structure.

Figure 3.21: EG1

	INTP	HR	PR	QRS	QT	QTca
1	NORMAL	50~beats/min	150~msec	88~msec	409~msec	389~mesc
2	NORMAL	56~beats/min	140~msec	100~msec	387~msec	365~mesc
3	NORMAL	55~beats/min	150~msec	91~msec	399~msec	376~mesc

Figure 3.22: EG2

NAME	COL1
HR	50~beats/min
PR	150~msec
QRS	88~msec
QT	409~msec
QTca	389~mesc
INTP	NORMAL
HR	56~beats/min
PR	140~msec
QRS	100~msec
QT	387~msec
QTca	365~mesc
INTP	NORMAL
HR	55~beats/min
PR	150~msec
QRS	91~msec
QT	399~msec
QTca	376~mesc
INTP	NORMAL

```
data EG3;
/*Specify length for standard variables*/
  length EGDTC EGTEST $20. EGTESTCD $8. EGORRESU $40. EGORRES EGSTRESC $100.;
  set EG2;
/*Derive EGTESTCD, EGSTRESC, EGORRES, EGCAT, EGPOS, EGORRESU, EGSTRESC */
  EGTESTCD=upcase(strip(_name_));     ❸
  EGTEST=put(EGTESTCD,$EGTESTCD.);
```

```
  if _name_="INTP" then do;                    ❹
    EGSTRESC=upcase(strip(scan(col1,1,"~")));
    if index(col1,"~") then EGORRES=upcase(strip(scan(col1,2,"~")));
    else EGORRES=EGSTRESC;
  end;

  else if _name_^="INTP" then do;
     if EGTESTCD="HR" then EGCAT="MEASUREMENT";
     else if EGTESTCD in ("PR","QRS","QT","QTCA") then EGCAT="INTERVAL";
     if EGTESTCD in ("HR","PR","QRS","QT","QTCA") then EGPOS="SUPINE";
     EGORRES=strip(scan(col1,1,"~"));
     EGORRESU=strip(scan(col1,2,"~"));
     EGSTRESC=EGORRES;
  end;
     /*Derive EGDTC*/
  EGD=datepart(EGDAT);
  EGM=timepart(EGDAT);
  EGDTC_DT=put(EGD,yymmdd10.);
  EGDTC_TM=put(EGM,tod8.);
  EGDTC=strip(EGDTC_DT)||"T"||strip(EGDTC_TM);    ❺
run;
```

❸ From Figure 3.22 and SDTM.EG specification, derive *EGTESTCD* from _name_, then use the put() function to create a user-defined format for *EGTEST.*

❹ In order to address all the different values of _name_, if its value is *INTP*, then use scan() to select the first string from *col1* as *EGSTRESC* and the second string from *COL1* as *EGORRES*. Otherwise, if _NAME_ is not equal with *INTP*, then use the scan() function to select the first string from *COL1* as *EGORRES* and the second string from *COL1* as *EGORRESU*. *EGSTRESC* is equal to *EGORRES*.

❺ Be diligent when you create variables with date and time, such as *EGDTC,* which is derived from the variable EGDAT (numeric) in the RAW.EG data set. First, use the datepart() function to select the date part of the EGDAT, then use the timepart() function to select the time part of *EGDAT*. Second, use the put() function with yymmdd10. to convert the date part value to character in yyyy-mm-dd format, then use the put() function with tod8. to switch the time part value to character in hh:mm:ss. Finally, concatenate date and time parts with "T" string.

To derive *EGBLFL*, we merge EX domain with EG3 to capture variable *EXSTDTC*, then convert the time character format (*EGDTC* and *EXSTDTC*) to time numeric format (*EGDTM* and *EXSTDTM*) to create these time variables.

```
/*Sort data set EG3 by USUBJID*/
proc sort data=EG3;
   by USUBJID;
run;
/*Sort data set SDTM.EX by USUBJID without duplicate values*/
proc sort data=SDTM.EX out=EX(keep=USUBJID EXSTDTC) nodupkey;
   by USUBJID;
run;
data EG4;
```

```
   /*Merge data set EG3 and EX*/
   merge EG3(in=a) EX(in=b keep=USUBJID EXSTDTC);
   by USUBJID;
   /*Derive EGDTM and EXSTDTM*/
   if length(EGDTC)>=10 then EGDTM=input(substr(EGDTC,1,16),e8601dt.);
   if length(EXSTDTC)>=10 then EXSTDTM=input(substr(EXSTDTC,1,16),e8601dt.);
run;

/*Select the condition when EGDTM is on or before EXSTDTM and no-missing
EGSTRESC to create the baseline flag with a value of "Y" for the last
nonmissing result for each EGTESTCD.*/
/*Sort data set EG4 by USUBJID, EGTESTCD,EGDTC */
proc sort data=EG4 out=BASEFL1;
   by USUBJID EGTESTCD EGDTC;
   where (EGDTM<=EXSTDTM and ^missing(EGSTRESC));
run;

data BASEFL2;
   set BASEFL1;
   /*Derive EGBLFL*/
   by USUBJID EGTESTCD EGDTC ;
   if last.EGTESTCD then EGBLFL="Y";
   keep USUBJID EGTESTCD EGDTC EGBLFL;
run;
```

Merge the BASELINE flag from the BASEFL2 data set to the EG4 data set, then merge with SDTM. TV to map the VISITNUM variable.

```
/*Sort data set BASEFL2 by USUBJID, EGTESTCD, EGDTC */
proc sort data=BASEFL2;
   by USUBJID EGTESTCD EGDTC;
run;
/*Sort data set EG4 by USUBJID, EGTESTCD, EGDTC */
proc sort data=EG4;
   by USUBJID EGTESTCD EGDTC;
run;
data EG5;
/*Merge data set EG4 and BASEFL2*/
   merge EG4(in=a) BASEFL2(in=b);
   by USUBJID EGTESTCD EGDTC;
   if a;
run;
proc sort data=EG5;
   by VISIT;
run;
proc sort data=SDTM.TV out=TV(keep=VISIT VISITNUM) nodupkey;
   by VISIT;
run;
data EG6;
   /*Merge data set EG5 and TV*/
   merge EG5(in=a) TV(in=b);
   by VISIT;
```

```
    if a;
run;
```

The code below creates the EGSEQ variable, which is used to uniquely record the sequence number of a subject within a domain. Please refer to the detailed explanation for the DSSEQ variable under DS domain in Section 3.6.2.

```
 /*Sort data set EG6 by USUBJID, EGCAT, EGTESTCD, EGDTC, VISIT */

proc sort data=EG6; by USUBJID EGCAT EGTESTCD EGDTC VISIT ; run;
data Final;
  set EG6;
/*Derive EGSEQ*/
  by USUBJID EGCAT EGTESTCD EGDTC VISIT;
  if first.USUBJID then EGSEQ = 0;
    EGSEQ+1;
    output;
  format _all_;
  informat _all_;
run;

libname SDTM ".../directory";
data SDTM.EG(label="Electrocardiogram Test Results");
/*Assign variable attributes such as label and length to conform with SDTM.EG
Specification (these will also be the same attributes as the SDTM IG).*/

attrib
  STUDYID      label = "Study Identifier"                      length = $20
  DOMAIN       label = "Domain Abbreviation"                   length = $2
  USUBJID      label = "Unique Subject Identifier"             length = $40
  EGSEQ        label = "Sequence Number"                       length = 8
  EGREFID      label = "ECG Reference ID"                      length = $20
  EGTESTCD     label = "ECG Test or Examination Short Name"    length = $20
  EGTEST       label = "ECG Test or Examination Name"          length = $20
  EGCAT        label = "Category for ECG"                      length = $20
  EGPOS        label = "ECG POSITION OF SUBJECTS"              length = $20
  EGORRES      label = "Result or Finding in Original Units"   length = $100
  EGORRESU     label = "Original Units"                        length = $40
  EGSTRESC     label = "Character Result/Finding in Std Format" length = $100
  EGMETHOD     label = "Method of ECG Test"                    length = $40
  EGBLFL       label = "Baseline Flag"                         length = $60
  VISIT        label = "Visit Name"                            length = $40
  VISITNUM     label = "Visit Number"                          length = 8
  EGDTC        label = "Date/Time of ECG"                      length = $40

      ;
  set Final;
   keep STUDYID DOMAIN USUBJID EGSEQ EGREFID EGTESTCD EGTEST EGCAT EGPOS
EGORRES EGORRESU EGSTRESC EGMETHOD EGBLFL VISIT VISITNUM EGDTC
      ;
run;
```

For the SDTM.EG domain, we show only the first 18 records out of 180 records for ease of viewing in Figure 3.23.

The final EG data set is saved in the SDTM library, as shown in Figure 3.23.

Figure 3.23: SDTM.EG

	STUDYID	DOMAIN	USUBJID	EGSEQ	EGREFID	EGTESTCD	EGTEST
1	ABC-001	EG	ABC-001-001-001	1 1		INTP	Interpretation
2	ABC-001	EG	ABC-001-001-001	2 1		INTP	Interpretation
3	ABC-001	EG	ABC-001-001-001	3 1		INTP	Interpretation
4	ABC-001	EG	ABC-001-001-001	4 1		PR	PR Interval
5	ABC-001	EG	ABC-001-001-001	5 1		PR	PR Interval
6	ABC-001	EG	ABC-001-001-001	6 1		PR	PR Interval
7	ABC-001	EG	ABC-001-001-001	7 1		QRS	QRS Interval
8	ABC-001	EG	ABC-001-001-001	8 1		QRS	QRS Interval
9	ABC-001	EG	ABC-001-001-001	9 1		QRS	QRS Interval
10	ABC-001	EG	ABC-001-001-001	10 1		QT	QT Interval
11	ABC-001	EG	ABC-001-001-001	11 1		QT	QT Interval
12	ABC-001	EG	ABC-001-001-001	12 1		QT	QT Interval
13	ABC-001	EG	ABC-001-001-001	13 1		QTCA	QTCA
14	ABC-001	EG	ABC-001-001-001	14 1		QTCA	QTCA
15	ABC-001	EG	ABC-001-001-001	15 1		QTCA	QTCA
16	ABC-001	EG	ABC-001-001-001	16 1		HR	Heart Rate
17	ABC-001	EG	ABC-001-001-001	17 1		HR	Heart Rate
18	ABC-001	EG	ABC-001-001-001	18 1		HR	Heart Rate

EGCAT	EGPOS	EGORRES	EGORRESU	EGSTRESC
		NORMAL		NORMAL
		NORMAL		NORMAL
		NORMAL		NORMAL
INTERVAL	SUPINE	150	msec	150
INTERVAL	SUPINE	140	msec	140
INTERVAL	SUPINE	150	msec	150
INTERVAL	SUPINE	88	msec	88
INTERVAL	SUPINE	100	msec	100
INTERVAL	SUPINE	91	msec	91
INTERVAL	SUPINE	409	msec	409
INTERVAL	SUPINE	387	msec	387
INTERVAL	SUPINE	399	msec	399
INTERVAL	SUPINE	389	mesc	389
INTERVAL	SUPINE	365	mesc	365
INTERVAL	SUPINE	375	mesc	375
MEASUREMENT	SUPINE	50	beats/min	50
MEASUREMENT	SUPINE	56	beats/min	56
MEASUREMENT	SUPINE	55	beats/min	55

EGMETHOD	EGBLFL	VISIT	VISITNUM	EGDTC
12 LEAD STANDARD	Y	DAY 1	3	2021-01-02T09:30:00
12 LEAD STANDARD		DAY 2	4	2021-01-03T09:30:00
12 LEAD STANDARD		DAY 3	5	2021-01-04T09:30:00
12 LEAD STANDARD	Y	DAY 1	3	2021-01-02T09:30:00
12 LEAD STANDARD		DAY 2	4	2021-01-03T09:30:00
12 LEAD STANDARD		DAY 3	5	2021-01-04T09:30:00
12 LEAD STANDARD	Y	DAY 1	3	2021-01-02T09:30:00
12 LEAD STANDARD		DAY 2	4	2021-01-03T09:30:00
12 LEAD STANDARD		DAY 3	5	2021-01-04T09:30:00
12 LEAD STANDARD	Y	DAY 1	3	2021-01-02T09:30:00
12 LEAD STANDARD		DAY 2	4	2021-01-03T09:30:00
12 LEAD STANDARD		DAY 3	5	2021-01-04T09:30:00
12 LEAD STANDARD	Y	DAY 1	3	2021-01-02T09:30:00
12 LEAD STANDARD		DAY 2	4	2021-01-03T09:30:00
12 LEAD STANDARD		DAY 3	5	2021-01-04T09:30:00
12 LEAD STANDARD	Y	DAY 1	3	2021-01-02T09:30:00
12 LEAD STANDARD		DAY 2	4	2021-01-03T09:30:00
12 LEAD STANDARD		DAY 3	5	2021-01-04T09:30:00

3.6.7 Laboratory Test Results (LB)

The laboratory test data includes tests and measurements performed on collected biological specimens. Laboratory test findings include, but are not limited to hematology, clinical chemistry, and urinalysis specimens.

Structure of LB

The structure of LB consists of Identifier, Timing, Topic, and Qualifier variables, as follows.

Identifier Variables:

- STUDYID: Required variable in character format, the unique study identifier. Each study has a unique study ID. In the example study described in this book, we populate it with "ABC-001".
- DOMAIN: Required variable in character format, compliant with SDTM Implementation Guide *DOMAIN* sets to "LB".
- USUBJID: Required variable in character format used to uniquely identify a subject across all studies. The USUBJID variable is derived using: STUDYID- SITED-SUBJID, such as "ABC-001-001-001".
- LBSEQ: Required variable in numeric format, the sequence number given to ensure uniqueness of subject records within a domain.
- LBSPID: Permissible variable in character format, the sponsor-defined identifier.

Timing Variables:

- LBDTC: Required variable in character format, the start date and time of Specification collection.
- VISIT: Permissible variable in character format, the Visit Name.
- VISITNUM: Expected variable in numeric format, the Visit Number.

Topic Variables:

- LBTESTCD: Required variable in character format, the short name of the measurement, test, or examination described in LBTEST.

Qualifier Variables:

- LBTEST: Required variable in character format, verbatim name of the test or examination used to obtain the measurement or finding. The value of LBTEST should not be greater than 40 characters.
- LBCAT: Expected variable in character format used to define a category of related records across subjects. For example: URINE ANALYSIS, CHEMISTRY, HEMATOLOGY.
- LBORRES: Expected variable in character format, the result of the measurement or finding as originally received or collected.
- LBORRESU: Expected variable in character format, the original units in which the data were collected. The unit of LBORRES.
- LBORNRLO: Expected variable in character format, the reference range lower limit for continuous measurement in original units. Should be populated only for continuous measurement.
- LBORNRHI: Expected variable in character format, the reference range higher limit for continuous measurement in original units. Should be populated only for continuous measurement.
- LBSTRESC: Expected variable in character format, the value is from LBORRES in a standard unit. LBSTRESC stores all character results. If results are numeric, then it should be stored in the numerical format under LBSTRESN.
- LBSTRESN: Expected variable in numeric format, the continuous or numeric result in standard format.
- LBSTRESU: Expected variable in character format, the standardized unit used for LBSTRESC or LBSTRESN.
- LBSTNRLO: Expected variable in character format, the lower end of reference range for continuous measurement for LBSTRESC/LBSTRESN in standardized units. Only populate for continuous results.
- LBSTRNRHI: Expected variable in character format, the higher end of reference range for continuous measurement for LBSTRESC/LBSTRESN in standardized units. Only populate for continuous results.
- LBNRIND: Expected variable in character format, the indication of where the value falls with respect to reference range defined by LBORNRLO and LBORNRHI, LBSTNRLO and LBSTNRHI.
- LBBLFL: Expected variable in character format, the identification of a baseline value.
- LBNAM: Permissible variable in character format, the name of the laboratory that performed the test.

LB eCRF Reference

Laboratory test findings are collected in each stage including "Screening", "Treatment", and "Follow-Up". The laboratory test data includes tests and measurements performed on collected biological specimens.

In terms of the CDISC CRF, there are two types of LB: Central Reading (Central) LB and Local Reading (Local) LB, and each are captured from a different source. For the Local LB, results are captured by a local vendor and the results can be recorded directly in the CRF.

Note: This example CRF is not shown in a vertical structure because the lab unit list is too lengthy. In the vertical structure, the unit codelist has to be inclusive of all analytes. However, we provide the LB test codelist for the Alkaline Phosphate and Calcium based on collected information from the CRF.

For the Central LB, the results are captured directly by a central vendor with an electronic device, the results are not recorded directly in the CRF. For the Local LB, results are captured by a local vendor and the results can be recorded directly in the CRF, as shown in Figure 3.24.

Figure 3.24: SDTM Annotated Laboratory Test Data (LB)

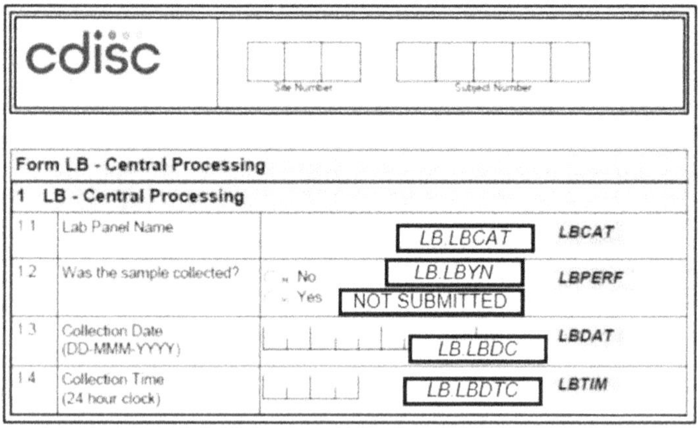

cdisc — Site Number — Subject Number

Form LB - Central Processing

1 LB - Central Processing

1.1	Lab Panel Name		LB.LBCAT	LBCAT
1.2	Was the sample collected?	No / Yes	LB.LBYN / NOT SUBMITTED	LBPERF
1.3	Collection Date (DD-MMM-YYYY)		LB.LBDC	LBDAT
1.4	Collection Time (24 hour clock)		LB.LBDTC	LBTIM

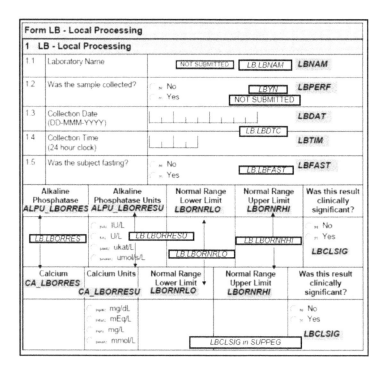

Figure 3.25: LBTEST CODELIST

LBTEST	LBTESTCD	LBCAT
Alkaline Phosphatase	ALP	CHEMISTRY
Calcium	CA	CHEMISTRY

LB Specification

Table 3.8: LB Specification

Variable	Label	Type	Length	Control Terminology	Origin	Notes
STUDYID	Study Identifier	char	20		Protocol: Set to "ABC-001"	This information that programmer can find in multiple raw data sets, such as RAW.IC; RAW.DM; RAW.AE, etc.
DOMAIN	Domain Abbreviation	char	2	DOMAIN	Assigned: Set to "LB"	

(Continued)

Table 3.8: (*Continued*)

Variable	Label	Type	Length	Control Terminology	Origin	Notes
USUBJID	Unique Subject Identifier	char	40		Derived: Concatenate among three variables below: STUDYID-SITED-SUBJID	
LBSEQ	Sequence Number	num	8		Derived: Sorted by USUBJID, LBCAT, LBTESTCD, LBDTC, VISITNUM, LBTPTNUM, LBSPID	
LBTESTCD	Lab Test or Examination Short Name	char	40			Raw LB.LBTESTCD
LBTEST	Lab Test or Examination Name	char	40		Assigned: Map from codelist	Equal to LB_CONVERTION. LBTEST after merge with LB_CONVERTION by LBTESTCD
LBCAT	Category of Lab Test	char	40			Raw LB.LBCAT
LBORRES	Result or Finding in Original Units	char	20			Uppercase of LBORRES.
LBORRESU	Original Units	char	40		Assigned	Raw LB.LBORRESU
LBORNRLO	Reference Range Lower Limit in Orig Unit	char	40			Raw LB.LBORNRLO

(*Continued*)

Table 3.8: (*Continued*)

Variable	Label	Type	Length	Control Terminology	Origin	Notes
LBORNRHI	Reference Range Upper Limit in Orig Unit	Char	40		Assigned	Raw LB.LBORNRHI
LBSTRESC	Character Result/ Finding in Std Format	Char	40		Assigned	LBSTRESC is calculated by multiplying the original result using conversion factor to convert to standard units. For example: LBSTRESC=LBORRES * CONVFAC. For conditional results, (>=, >, <=, <) the condition is retained and conversion factor is only applied to the numeric part.
LBSTRESN	Numeric Result/ Finding in Standard Units	num	8			For numeric result, LBSTRESN is calculated: LBSTRESN = LBORRES * CONVFAC
LBSTRESU	Standard Units	char	40		Assigned	Merge the standard units from the CONVFAV data set, merging by LBCAT, LBTESTCD, and LBORRESU.
LBSTNRLO	Reference Range Lower Limit- Std Units	num	8		Designed:	LBSTNRLO= LBORNRLO * CONVFAC for numeric values
LBSTNRHI	Reference Range Upper Limit- Std Units	num	8		Designed:	LBSTNRHI=LBORNRHI * CONVFAC for numeric values

(*Continued*)

Table 3.8: (*Continued*)

Variable	Label	Type	Length	Control Terminology	Origin	Notes
LBBLFL	Baseline Flag	char	2			Indicator used to identify a baseline value. Set to "Y" for the last nonmissing result for each LB test that occurred before first study dose
LBDTC	Date/Time of Specimen Collection	Char	40			Raw LB.LBDAT
VISIT	Visit Name	char	40			Raw LB.EVENT
VISITNUM	Visit Number	num	8			Map VISITNUM from SDTM.TV

LB Programming

In order to create SDTM.LB, we first run a PROC CONTENTS on the raw data. Please see the code below.

Note: For every domain, the corresponding SAS name is domain.sas. For example, LB is created by lb.sas, so all code in this section belongs to lb.sas.

```
/* Begin writing SAS program lb.sas*/
/*show structure of the raw LB dataset*/
proc contents data=RAW.LB;
run;
```

First, as seen in Figure 3.26, we use PROC CONTENTS to see that there are 23 variables in the raw LB data set. Further, we can see the Type, Length (Len), Format, and Informat of each variable. These variable attributes are assigned by SAS based on the raw data format. The Label is assigned in SAS based on the first row of the raw LB data set in the Excel sheet.

Note: The purpose of running PROC CONTENTS is to get a clear idea of what the raw metadata looks like and compare how similar it is to the SDTM domain that we will create from it.

The code below creates the variables DOMAIN, USUBJID, LBDTC, LBNRIND, LBNAM, LBFAST, LBORRES, LBORRESU, LBORNRLO, and LBORNRHI.

Figure 3.26: Alphabetic List of Variables and Attributes

Alphabetic List of Variables and Attributes

#	Variable	Type	Len	Format	Informat	Label
2	COUNTRY	Char	13	$13	$13	COUNTRY
12	LBCAT	Char	10	$10	$10	LBCAT
23	LBCLSIG	Char	1	$1	$1	LBCLSIG
7	LBDAT	Num	8	DATE9		LBDAT
9	LBFAST	Char	1	$1	$1	LBFAST
5	LBNAM	Char	20	$20	$20	LBNAM
15	LBNRIND	Char	6	$6	$6	LBNRIND
17	LBORNRHI	Num	8	BEST		LBORNRHI
16	LBORNRLO	Num	8	BEST		LBORNRLO
13	LBORRES	Num	8	BEST		LBORRES
14	LBORRESU	Char	5	$5	$5	LBORRESU
22	LBSTNRHI	Num	8	BEST		LBSTNRHI
21	LBSTNRLO	Num	8	BEST		LBSTNRLO
18	LBSTRESC	Num	8	BEST		LBSTRESC
19	LBSTRESN	Num	8	BEST		LBSTRESN
20	LBSTRESU	Char	5	$5	$5	LBSTRESU
10	LBTEST	Char	20	$20	$20	LBTEST
11	LBTESTCD	Char	3	$3	$3	LBTESTCD
8	LBTIM	Num	8	TIME		LBTIM
6	LBYN	Char	1	$1	$1	LBYN
3	SITEID	Char	3	$3	$3	SITEID
1	STUDYID	Char	7	$7	$7	STUDYID
4	SUBJID	Char	3	$3	$3	SUBJID

```
data LB1;
  /*Specify length for standard variables*/
  length USUBJID VISIT $40.;

  /*Rename STUDYID, LBNAM, LBNRIND, LBFAST, LBORRES, LBORNRLO, LBORNRHI */
  set RAW.LB(where=(LBYN="Y") rename=(STUDYID=STUDYID_ LBNRIND=LBNRIND_
LBFAST=LBFAST_ LBORRES=LBORRES_ LBORNRLO=LBORNRLO_ LBORNRHI=LBORNRHI_ ));

  /*Define DOMAIN*/
  DOMAIN="LB";

  /*Derive STUDYID, USUBJID, LBDTC, LBNRIND, LBFAST, LBORRES, LBORRESU,
LBORNRLO, LBORNRHI */
  STUDYID=strip(STUDYID_);
  USUBJID=STRIP(STUDYID)||"-"||STRIP(SITEID)||"-"||STRIP(SUBJID);

  if ^missing(LBDAT) and ^missing(LBTIM) then
LBDTC=put(LBDAT,yymmdd10.)||"T"||put(LBTIM,tod5.);        ❶
  else if ^missing(LBDAT) and missing(LBTIM) then
LBDTC=strip(put(LBDAT,yymmdd10.));

  else if missing(LBDAT) and ^missing(LBTIM) then LBDTC="-----T"
||put(LBTIM,tod5.);
```

```
    LBNRIND=upcase(strip(LBNRIND_));
    LBFAST=upcase(strip(LBFAST_));
    LBORRES=strip(put(LBORRES_,best.));
    LBORRESU=upcase(strip(LBORRESU));
    LBORNRLO=strip(put(LBORNRLO_,best.));
    LBORNRHI=strip(put(LBORNRHI_,best.));

    if index(upcase(EVENT),"SCREENING") then VISIT="SCREENING";
    else if index(upcase(EVENT),"FOLLOW UP") then VISIT="FOLLOW-UP";
    else if index(upcase(EVENT),"DAY") then VISIT="DAY
"||strip(substr(upcase(EVENT),length(EVENT)-1));
    else VISIT=strip(upcase(EVENT));

run;
```

❶ Be diligent when you create variables with date and time, such as *LBDTC,* which is derived from the variable *LBDAT* (numeric) in the RAW.LB data set. First, use the datepart() function to select the date part of the *LBDAT,* then use the timepart() function to select the time part of *LBDAT.* Second, use the put() function with yymmdd10. to convert the date part value to character in yyyy-mm-dd format, then use the put() function with tod8. to switch the time part value to character in hh:mm:ss. Finally, concatenate date and time parts with "T" string.

Different laboratories might have different original lab test units. In order to provide accurate and convenient laboratory results to FDA, we need to convert the original lab test units to the standard units appropriate to the lab test. RAW.LB_CONVERSION includes all possible original units and standard units for each *LBTEST* under *LBCAT,* as well as the corresponding conversion factor.

Note: For RAW.LB conversion, this data is typically created over time, with new LBTEST parameters being added cumulatively. This could vary by name or even process for conversion, depending on the sponsor.

Therefore, we use the following variables, including *ORIGINAL_UNIT, STANDARD_UNIT, CONVERSION* from *RAW.LB_CONVERSION,* and merge with LB2 data set using the following variables: *LBCAT, LBTESTCD, ORIGNIAL_UNIT.*

```
/*pull the ORIGINAL_UNIT, STANDARD_UNIT,CONVERSION from RAW.LB_CONVERSION*/
proc sql noprint;
 create table LB2 as
   select a.*, b.ORIGINAL_UNIT, b.STANDARD_UNIT, b.CONVERSION from LB1 as a
   left join RAW.LB_CONVERTION as b
on a.LBCAT=b.LBCAT and a.LBTESTCD=b.LBTESTCD and a.LBORRESU=b.ORIGINAL_UNIT
 order by STUDYID, USUBJID, LBCAT, LBTESTCD, LBDTC
      ;
quit;
```

The code below creates the variables LBSTRESN, LBSTRESC, LBSTRNHI, LBSTNRLO, LBTEST, and LBTESTCD.

```
data LB3;
```

```
/*Specify length for standard variables*/
length LBSTRESN LBSTNRLO LBSTNRHI 8 LBSTRESU LBSTRESC LBTEST LBTESTCD $40.;
/*Rename LBSTRESC, LBSTRESN, LBSTRESU, LBSTNRLO, LBSTNRHI, LBTESTCD,
LBTEST */

set LB2(rename=(LBSTRESC=LBSTRESC_ LBSTRESN=LBSTRESN_ LBSTRESU=LBSTRESU_
LBSTNRLO=LBSTNRLO_ LBSTNRHI=LBSTNRHI_ LBTESTCD=LBTESTCD_ LBTEST=LBTEST_));

/*Derive LBSTRESN, LBSTRESC */
if substr(LBORRES,1,1)=">" then do;       ❷
  LBSTRESN=input(substr(LBORRES,2),best.)*CONVERSION;
  LBSTRESC=">"||strip(put(LBSTRESN,best.));
end;
if substr(LBORRES,1,1)="<" then do;
  LBSTRESN=input(substr(LBORRES,2),best.)*CONVERSION;
  LBSTRESC="<"||strip(put(LBSTRESN,best.));
end;
if substr(LBORRES,1,2)=">=" then do;
  LBSTRESN=input(substr(LBORRES,3),best.)*CONVERSION;
  LBSTRESC=">="||strip(put(LBSTRESN,best.));
end;
if substr(LBORRES,1,2)="<=" then do;
  LBSTRESN=input(substr(LBORRES,3),best.)*CONVERSION;
  LBSTRESC="<="||strip(put(LBSTRESN,best.));
end;
else if ^missing(LBORRES) then do;
  LBSTRESN=input(LBORRES,best.)*CONVERSION;
  LBSTRESC=strip(put(LBSTRESN,best.));
end;
/*Derive LBSTNRHI, LBSTNRLO, LBTEST, LBTESTCD */

LBSTNRHI=input(LBORNRHI,best.)*CONVERSION;
LBSTNRLO=input(LBORNRHI,best.)*CONVERSION;

if ^missing(CONVERSION) and ^missing(STANDARD_UNIT) then LBSTRESU=STANDARD_
UNIT;
LBTEST=upcase(strip(LBTEST_));
LBTESTCD=upcase(strip(LBTESTCD_));
run;
```

❷ The values of *LBORRES* contain special conditional symbols, such as ">=", "<=", ">", "<". We include only the numeric parts (removing the special symbols) then retain the numeric part of *LBORRES* and multiply the conversion factor with the numeric part for each lab test in order to derive *LBSTRESN*.

In order to create variable LBBLFL using a numeric format, we merge EX domain with EG3 to get the variable *EXSTDTC*, then convert the time character format (*LBDTC* and *EXSTDTC*) to the time numeric format (*LBDTM* and *EXSTDTM*).

```
proc sort data=LB3;
  by USUBJID;
```

```
run;
proc sort data=SDTM.EX out=EX(keep=USUBJID EXSTDTC) nodupkey;

   by USUBJID;
run;

data LB4;
/*Merge LB3 and EX*/
   merge LB3(in=a) EX(in=b keep=USUBJID EXSTDTC);
   by USUBJID;
/*Derive LBDTC, EXSTDTM*/
   if length(LBDTC)>=10 then LBDTM=input(substr(LBDTC,1,16),e8601dt.);
   if length(EXSTDTC)>=10 then EXSTDTM=input(substr(EXSTDTC,1,16),e8601dt.);
run;

/* Select the condition when LBDTM is on or before EXSTDTM and nonmissing
LBSTRESC, then create the baseline flag "Y" for the last nonmissing result
for each LBTESTCD.*/
/*Sort data set LB4 by USUBJID, LBTESTCD, LBDTC */

proc sort data=LB4 out=BASEFL1;
   by USUBJID LBTESTCD LBDTC;
   where (LBDTM<=EXSTDTM and ^missing(LBSTRESC));
run;

data BASEFL2;
   set BASEFL1;
   /*Derive LBBLFL*/
   by USUBJID LBTESTCD LBDTC ;

/*Derive LBBLFL*/
   if last.LBTESTCD then LBBLFL="Y";
   keep USUBJID LBTESTCD LBDTC LBBLFL;
run;
```

We merge BASELINE flag from BASEFL2 data set to LB4 data set and create variable *LBCLSIG* based on the CRF.

```
proc sort data=BASEFL2;
   by USUBJID LBTESTCD LBDTC;
run;

proc sort data=LB4;
   by USUBJID LBTESTCD LBDTC;
run;

data LB5;

/*Merge LB4 and BASEFL2*/
   merge LB4(in=a rename=(LBCLSIG=LBCLSIG_)) BASEFL2(in=b);
   by USUBJID LBTESTCD LBDTC;
   if a;
     if LBSTRESC="ABNORMAL" then do;
       if LBCLSIG_="No" then LBCLSIG="N";
       else if LBCLSIG_="Yes" then LBCLSIG="Y";        ❸
```

```
        end;
    else LBCLSIG="";

run;
```

❸ In this case, *LBCLSIG* is null since there is no "ABNORMAL" value in *LBSTRESC*.

```
/*Sort data set LB5 and SDTM.TV*/
proc sort data=LB5;
  by VISIT;
run;

proc sort data=SDTM.TV out=TV(keep=VISIT VISITNUM) nodupkey;
  by VISIT;

run;
/*Merge LB5 with TV to derive VISITNUM*/
data LB6;
  merge LB5(in=a) TV(in=b);
  by VISIT;
  if a;
run;
```

In the code below, we create the *LBSEQ* variable, which is used to uniquely record the sequence number of a subject within a domain. To create *LBSEQ,* please refer to the detailed explanation for the *DSSEQ* variable under the DS domain in Section 3.6.2.

```
/*Sort data set LB6 by USUBJID, LBTESTCD, LBDTC, VISIT*/
proc sort data=LB6;
    by USUBJID LBTESTCD LBDTC VISIT;
run;

data Final;
    set LB6;
    by USUBJID LBTESTCD LBDTC VISIT;
 /*Derive LBSEQ*/
  if first.USUBJID then LBSEQ = 0;
    LBSEQ+1;
    output;
  format _all_;
  informat _all_;
run;

libname SDTM "/directory";
data SDTM.LB(label="LABORATORY TEST Results");
/*Assign variable attributes such as label and length to conform with SDTM.LB
Specification (these will also be the same attributes as the SDTM IG).*/

attrib
  STUDYID      label = "Study Identifier"                      length = $20
  DOMAIN       label = "Domain Abbreviation"                   length = $2
  USUBJID      label = "Unique Subject Identifier"             length = $40
  LBSEQ        label = "Sequence Number"                       length = 8
  LBTESTCD     label = "Lab Test or Examination Short Name"    length = $40
```

```
LBTEST      label = "Lab Test or Examination Name"              length = $40
LBCAT       label = "Category for Lab Test"                     length = $40
LBORRES     label = "Result or Finding in Original Units"       length = $20
LBORRESU    label = "Original Units"                            length = $40
LBORNRLO    label = "Reference Range Lower Limit in Orig Unit"length = $40
LBORNRHI    label = "Reference Range Upper Limit in Orig Unit"length = $40
LBSTRESC    label = "Character Result/Finding in Std Format" length = $40
LBSTRESN    label = "Numeric Result/Finding in Standard Units"length = 8
LBSTRESU    label = "Standard Units"                            length = $40
LBSTNRLO    label = "Reference Range Lower Limit-Std Units"   length = 8
LBSTNRHI    label = "Reference Range Upper Limit-Std Units"   length = 8
LBNRIND     label = "Reference Range for Char Rslt-Std Units" length = $40
LBBLFL      label = "Baseline Flag"                             length = $2
LBDTC       label = "Date/Time of Specimen Collection"          length = $40
VISIT       label = "Visit Name"                                length = $40
VISITNUM    label = "Visit Number"                              length = 8

        ;

  set Final;

   keep STUDYID DOMAIN USUBJID LBSEQ LBTESTCD LBTEST LBCAT LBORRES LBORRE-
SU LBORNRLO LBORNRHI LBSTRESC LBSTRESN LBSTRESU LBSTNRLO LBSTNRHI LBNRIND
LBBLFL LBDTC VISIT VISITNUM

        ;

run;
```

The final LB data set is saved in the SDTM library, as shown in Figure 3.27.

Figure 3.27: SDTM.LB

	DOMAIN	USUBJID	LBSEQ	LBTESTCD	LBTEST	LBCAT
1	LB	ABC-001-001-001	1	ALP	ALKALINE PHOSPHATASE	CHEMISTRY
2	LB	ABC-001-001-001	2	CA	CALCIUM	CHEMISTRY
3	LB	ABC-001-001-002	1	ALP	ALKALINE PHOSPHATASE	CHEMISTRY
4	LB	ABC-001-001-002	2	CA	CALCIUM	CHEMISTRY
5	LB	ABC-001-001-003	1	ALP	ALKALINE PHOSPHATASE	CHEMISTRY
6	LB	ABC-001-001-003	2	CA	CALCIUM	CHEMISTRY
7	LB	ABC-001-001-004	1	ALP	ALKALINE PHOSPHATASE	CHEMISTRY
8	LB	ABC-001-001-004	2	CA	CALCIUM	CHEMISTRY
9	LB	ABC-001-001-005	1	ALP	ALKALINE PHOSPHATASE	CHEMISTRY
10	LB	ABC-001-001-005	2	CA	CALCIUM	CHEMISTRY
11	LB	ABC-001-001-006	1	ALP	ALKALINE PHOSPHATASE	CHEMISTRY
12	LB	ABC-001-001-006	2	CA	CALCIUM	CHEMISTRY
13	LB	ABC-001-001-007	1	ALP	ALKALINE PHOSPHATASE	CHEMISTRY
14	LB	ABC-001-001-007	2	CA	CALCIUM	CHEMISTRY
15	LB	ABC-001-001-008	1	ALP	ALKALINE PHOSPHATASE	CHEMISTRY
16	LB	ABC-001-001-008	2	CA	CALCIUM	CHEMISTRY
17	LB	ABC-001-001-009	1	ALP	ALKALINE PHOSPHATASE	CHEMISTRY
18	LB	ABC-001-001-009	2	CA	CALCIUM	CHEMISTRY
19	LB	ABC-001-001-010	1	ALP	ALKALINE PHOSPHATASE	CHEMISTRY
20	LB	ABC-001-001-010	2	CA	CALCIUM	CHEMISTRY

LBORRES	LBORRESU	LBORNRLO	LBORNRHI	LBSTRESC	LBSTRESN
62	U/L	50	100	62	62
5	MG/DL	8.5	10	2.25	2.25
65	U/L	50	100	65	65
5.5	MG/DL	8.5	10	2.375	2.375
66	U/L	50	100	66	66
9.2	MG/DL	8.5	10	2.3	2.3
67	U/L	50	100	67	67
9.4	MG/DL	8.5	10	2.35	2.35
64	U/L	50	100	64	64
9.5	MG/DL	8.5	10	2.375	2.375
63	U/L	50	100	63	63
9.2	MG/DL	8.5	10	2.3	2.3
66	U/L	50	100	66	66
9.3	MG/DL	8.5	10	2.325	2.325
60	U/L	50	100	60	60
9	MG/DL	8.5	10	2.25	2.25
62	U/L	50	100	62	62
9.2	MG/DL	8.5	10	2.3	2.3
63	U/L	50	100	63	63
9.2	MG/DL	8.5	10	2.3	2.3

LBSTRESU	LBSTNRLO	LBSTNRHI	LBNRIND	LBBLFL	LBDTC	VISIT	VISITNUM
U/L	100	100	NORMAL	Y	2021-01-02T09:45	DAY 1	3
mmol/L	2.5	2.5	NORMAL	Y	2021-01-02T09:45	DAY 1	3
U/L	100	100	NORMAL	Y	2021-01-02T09:45	DAY 1	3
mmol/L	2.5	2.5	NORMAL	Y	2021-01-02T09:45	DAY 1	3
U/L	100	100	NORMAL	Y	2021-01-02T09:45	DAY 1	3
mmol/L	2.5	2.5	NORMAL	Y	2021-01-02T09:45	DAY 1	3
U/L	100	100	NORMAL	Y	2021-01-02T09:45	DAY 1	3
mmol/L	2.5	2.5	NORMAL	Y	2021-01-02T09:45	DAY 1	3
U/L	100	100	NORMAL	Y	2021-01-02T09:45	DAY 1	3
mmol/L	2.5	2.5	NORMAL	Y	2021-01-02T09:45	DAY 1	3
U/L	100	100	NORMAL	Y	2021-01-02T09:45	DAY 1	3
mmol/L	2.5	2.5	NORMAL	Y	2021-01-02T09:45	DAY 1	3
U/L	100	100	NORMAL	Y	2021-01-02T09:45	DAY 1	3
mmol/L	2.5	2.5	NORMAL	Y	2021-01-02T09:45	DAY 1	3
U/L	100	100	NORMAL	Y	2021-01-02T09:45	DAY 1	3
mmol/L	2.5	2.5	NORMAL	Y	2021-01-02T09:45	DAY 1	3
U/L	100	100	NORMAL	Y	2021-01-02T09:45	DAY 1	3
mmol/L	2.5	2.5	NORMAL	Y	2021-01-02T09:45	DAY 1	3
U/L	100	100	NORMAL	Y	2021-01-02T09:45	DAY 1	3
mmol/L	2.5	2.5	NORMAL	Y	2021-01-02T09:45	DAY 1	3

3.7 Trial Design Domains

The design of a clinical trial is to plan subjects' experience during the course of the trial, including milestones such as planned visits, assessments, etc. as well as what data will be collected during the trial.

Note: There are many combinations of clinical trial designs with different treatments, study milestones, and trial information. For the purposes of this book, we describe only the *structure* and *intent* of each trial design domain as well as a simple example of each trial design data set.

This section includes the descriptions and programming practices for the five most common and essential Trial Domains: Trial Summary (TS), Trial Arm (TA), Trial Element (TE), Trial Visit (TV), and Trial Inclusion/Exclusion (TI).

Trial Design data sets can usually be created from the protocol and are used in the eCRF design to ensure all data points related to the Trial Design are captured.

3.7.1 Trial Summary Data Set (TS)

The Trial Summary (TS) data set includes important detailed information about the clinical trial, including the protocol title, objective, indication, and so on. The structure of the Trial Summary data set is one record per parameter per occurrence.

Structure of Trial Summary (TS)

The structure of TS consists of Identifier, Topic, and Qualifier variables, as follows.

Identifier Variables:

- STUDYID: Required variable in character format, the unique study identifier. Each study has a unique study ID. In the example study described in this book, we populate it with "ABC-001".
- DOMAIN: Required variable in character format, compliant with SDTM Implementation Guide. *DOMAIN* sets to "TS".
- TSSEQ: Required variable in numeric format, the sequence number is given to ensure uniqueness within a subject. It is used as a key to distinguish between multiple occurrences of the same parameter. For example, each dose would be in a separate record.

Topic Variables:

- TSPARMCD: Required variable in numeric format, the trial summary parameter short name.

Qualifier Variables:

- TSPARM: Required variable in numeric format, the trial summary parameter.
- TSVAL: Required variable in numeric format, the parameter value.
- TSVALNF: Permissible variable in numeric format, the parameter null flavor. This is the code of the term in TSVAL, which is a coded value that offers additional information that cannot be filled in the *TSVAL*.
- TSVCDREF: Expected variable in numeric format, the name of the reference terminology.

TS Specification

Table 3.9: TS Specification

Variable	Label	Type	Length	Control Terminology	Origin	Notes
STUDYID	Study Identifier	char	20		Protocol: Set to "ABC-001"	

(Continued)

Table 3.9: (*Continued*)

Variable	Label	Type	Length	Control Terminology	Origin	Notes
DOMAIN	Domain Abbreviation	char	2	DOMAIN	Assigned: Set to "TS"	
TSSEQ	Sequence Number	num	8		Assigned	
TSPARMCD	Trial Summary Parameter Short Name	char	8		Assigned	
TSPARM	Trial Summary Parameter	char	8		Assigned	
TSVAL	Parameter Value	char	40		Assigned	
TSVALFN	Parameter Null Flavor	char	100		Assigned	
TSVCDREF	Name of the Reference Terminology	char	20		Assigned	

SDTM.TS

Figure 3.28: SDTM.TS

	STUDYID	DOMAIN	TSSEQ	TSPARMCD	TSPARM	TSVAL	TSVAL	TSVCDREF
1	ABC-001	TS	1	ACTSUB	Actual Number of Subjects	15		
2	ABC-001	TS	1	ADAPT	Adaptive Design	N		CDISC
3	ABC-001	TS	1	ADDON	Added on to Existing Treatments	N		CDISC
4	ABC-001	TS	1	AGEMAX	Planned Maximum Age of Subjects	P50Y		ISO 8601
5	ABC-001	TS	1	AGEMIN	Planned Minimum Age of Subjects	P18Y		ISO 8601
6	ABC-001	TS	1	DCUTDESC	Data Cutoff Description	Final Database Lock		
7	ABC-001	TS	1	DCUTDTC	Data Cutoff Date	2021-01-21		ISO 8601
8	ABC-001	TS	1	DOSE	Dose per Administration			
9	ABC-001	TS	1	DOSFRQ	Dosing Frequency	UNKNOWN		CDISC
10	ABC-001	TS	1	FCNTRY	Planned Country of Investigational Sites	USA		ISO 3166
11	ABC-001	TS	1	HLTSUBJI	Healthy Subject Indicator	N		CDISC
12	ABC-001	TS	1	INDIC	Trial Disease/Condition Indication	Acute pain		SNOMED
13	ABC-001	TS	1	INTMODEL	Intervention Model	PARALLEL		CDISC
14	ABC-001	TS	1	INTTYPE	Intervention Type	SURGERY		CDISC
15	ABC-001	TS	1	LENGTH	Trial Length	P10D		ISO 8601
16	ABC-001	TS	1	NARMS	Planned Number of Arms	1		
17	ABC-001	TS	1	OBJPRIM	Trial Primary Objective	To evaluate the efficacy of Drug A in subjects who suffer with Pain		
18	ABC-001	TS	1	OBJSEC	Trial Secondary Objective	To evaluate the safety of Drug A in subjects who suffer with Pain		
19	ABC-001	TS	1	OUTMSPRI	Primary Outcome Measure	SEIZURE FREQUENCY		
20	ABC-001	TS	1	OUTMSSEC	Secondary Outcome Measure	SEIZURE FREQUENCY		
21	ABC-001	TS	1	OUTMSADD	Additional Outcome Measure	SEIZURE FREQUENCY		
22	ABC-001	TS	1	PCLAS	Pharmacologic Class		NAV	ISO 21090
23	ABC-001	TS	1	PLANSUB	Planned Number of Subjects	10		
24	ABC-001	TS	1	RANDQT	Randomization Quotient	N		
25	ABC-001	TS	1	RANDOM	Trial is Randomized	N		CDISC
26	ABC-001	TS	1	REGID	Registry Identifier	NCT123456789		ClinicalTrials.GOV
27	ABC-001	TS	1	SENDTC	Study End Date	2021-01-20		EudraCT
28	ABC-001	TS	1	SEVCRIT	Severity Criteria	NA		ISO 8601
29	ABC-001	TS	1	SEXPOP	Sex of Participants	BOTH		
30	ABC-001	TS	1	SPONSOR	Clinical Study Sponsor	MALE		CDISC
31	ABC-001	TS	1	SSTDTC	Study Start Date	XX Pharma		
32	ABC-001	TS	1	STOPRULE	Study Stop Rules	NA		ISO 8601
33	ABC-001	TS	1	STRATFCT	Stratification Factor	NONE		
34	ABC-001	TS	1	STYPE	Study Type	PAIN		
35	ABC-001	TS	1	TBLIND	Trial Blinding Schema	OBSERVATIONAL		CDISC
36	ABC-001	TS	1	TCNTRL	Control Type	OPEN LABEL		CDISC
37	ABC-001	TS	1	TINDTP	Trial Intent Type	NA		SNOMED
38	ABC-001	TS	1	TITLE	Trial Title	TREATMENT		CDISC
39	ABC-001	TS	1	TPHASE	Trial Phase Classification	PHASE I TRIAL		
40	ABC-001	TS	1	TRT	Investigational Therapy or Treatment	PHASE I TRIAL		CDISC
41	ABC-001	TS	1	TTYPE	Trial Type	TREATMENT		

3.7.2 Trial Arm Data Set (TA)

The Trial Arm (TA) domain describes each planned arm in the trial. An arm is an ordered sequence of elements. (Elements will be obvious to the reader in Figure 3.29.) There are three related concepts. The Arm is a planned path through the trial. It comprises the entire time period of the trial. The Element is a basic block in the trial design. Lastly, the Branch describes the treatment, which in Figure 3.29 is "Drug A", "Drug B", and "Placebo". For most trials, the assignment to an Arm happens at a single time point; this trial has one branch point.

Figure 3.29: The Relationship among Element, Branch, and Arm

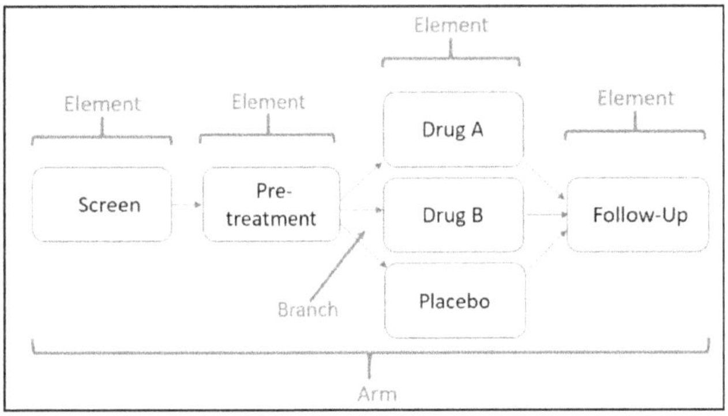

Structure of Trial Arm (TA)

The structure of TA consists of Identifier, Timing, Topic, Qualifier, and Rule variables, as follows.

Identifier Variables:

- STUDYID: Required variable in character format, the unique study identifier. Each study has a unique study ID. In the example study described in this book, we populate it with "ABC-001".
- DOMAIN: Required variable in character format, compliant with SDTM Implementation Guide. *DOMAIN* sets to "TA".

Timing Variables:

- TAETORD: Required variable in character format, the planned order of element within arm.
- EPOCH: Required variable in character format, the name of the trial epoch with which this element of the arm is associated.

Topic Variables:

- ARMCD: Required variable in character format, the planned arm code.

Qualifier Variables:

- ARM: Required variable in character format, the description of planned arm.
- ETCD: Required variable in character format, the element code.
- ELEMENT: Permissible variable in character format the description of element.

Rule Variables:

- TRBRANCH: Expected variable in character format, the decision points at the end of elements from which subjects "branch" into an element unique to their assigned ARM. For example, "Randomized to Drug A".
- TATRANS: Expected variable in character format, the transition rule, is used to describe the conditions under which subjects can skip one or more elements, and to which element would go to next.

TA Specification

Table 3.10: TA Specification

Variable	Label	Type	Length	Control Terminology	Origin	Notes
STUDYID	Study Identifier	char	20		Protocol: Set to "ABC-001"	
DOMAIN	Domain Abbreviation	char	2	DOMAIN	Assigned: Set to "TA"	
ARMCD	Planned Arm Code	char	20		Assigned	
ARM	Planned Arm	char	100		Protocol	
TAERORD	Planned Order of Element Within Arm	num	8		Protocol	

(Continued)

Table 3.10: (*Continued*)

Variable	Label	Type	Length	Control Terminology	Origin	Notes
ETCD	Element Code	char	100		Assigned	
ELEMENT	Description Element	char	100		Protocol	
TABRANCH	Branch	char	100		Protocol	
TATRANS	Transition Rule	char	100		Protocol	
EPOCH	Epoch	char	60		Assigned	

SDTM.TA

Figure 3.30: SDTM.TA

	STUDYID	DOMAIN	ARMCD	ARM	TAETORD	ETCD	ELEMENT
1	ABC-001	TA	DRUG A	DRUG A 10 mg	1	SCRE	Screening
2	ABC-001	TA	DRUG A	DRUG A 10 mg	2	PRE	Pretreatment
3	ABC-001	TA	DRUG A	DRUG A 10 mg	3	TRT	Treatment
4	ABC-001	TA	DRUG A	DRUG A 10 mg	4	SFU	Safety Follow Up

TABRANCH	TATRANS	EPOCH
		SCREENING
		PRETREATMENT
	if diease progression then go to Satey Follow Up Epoch	OPEN LABEL TREATMENT
		SAFETY FOLLOW UP

3.7.3 Trial Element Data Set (TE)

The Trial Element (TE) domain contains the definitions of the elements that appear in the Trial Element data set (refer to Figure 3.31. For example, some elements are the first day the subject takes treatment, the last day the subject takes treatment, and anytime there is a planned change of the treatment.

Structure of Trial Element (TE)

The structure of TE consists of Identifier, Topic, Qualifier, and Rule variables, as follows.

Identifier Variables:

- STUDYID: Required variable in character format, the unique study identifier. Each study has a unique study id. In the example study described in this book, we populate it with "ABC-001".
- DOMAIN: Required variable in character format, compliant with SDTM Implementation Guide DOMAIN sets to "TE".

Topic Variables:

- ETCD: Required variable in character format, the Element code, the companion to ELEMENT, is limited to 8 characters and does not have special character restrictions.

Qualifier Variables:

- ELEMENT: Required variable in character format, the description of Element.

Rule Variables:

- TESTRL: Required variable in character format, the rule for start of Element.
- TEENRL: Permissible variable in character format, the rule for end of Element.

TE Specification

Table 3.11: TE Specification

Variable	Label	Type	Length	Control Terminology	Origin	Notes
STUDYID	Study Identifier	char	20		Protocol: Set to "ABC-001"	
DOMAIN	Domain Abbreviation	char	2	DOMAIN	Assigned: Set to "TE"	
ETCD	Element Code	char	20		Assigned	
ELEMENT	Element	char	200		Protocol	
TESTRL	Rule for Start of Element	num	8		Protocol	
TEENRL	Rule for End of Element	char	200		Protocol	

SDTM.TE

Figure 3.31: SDTM.TE

	STUDYID	DOMAIN	ETCD	ELEMENT	TESTRL	TEENRL
1	ABC-001	TE	SCRE	Screening	Inform consent obtained	Admission to clinic
2	ABC-001	TE	PRE	Pretreatment	Admission to clinic	First dose of study drug, where drug is Drug A
3	ABC-001	TE	TRT	Treatment	First dose of study drug, where drug is Drug A	First dose of study drug on Day 1
4	ABC-001	TE	SFU	Safety Follow Up	48 hours after last dose	At Trial Exit

3.7.4 Trial Visit Data Set (TV)

The Trial Visit (TV) domain describes all the planned visits in a trial. Visits are defined as "clinical encounters" and are described using the time variables VISIT, VISITNUM, and VISITDY.

Note: Visits are a collection of activities that might include dispensation of study treatment and certain measurements such as vital signs and laboratory tests.

Structure of Trial Visit (TV)

The structure of TV consists of Identifier, Topic, Qualifier, and Rule variables, as follows.

Identifier Variables:

- STUDYID: Required variable in character format, the unique study identifier. Each study has a unique study id. In the example study described in this book, we populate it with "ABC-001".
- DOMAIN: Required variable in character format, compliant with SDTM Implementation Guide. *DOMAIN* sets to "TV".

Topic Variables:

- VISITNUM: Required variable in character format, the clinical visit number using for sorting.

Qualifier Variables:

- VISIT: Permissible variable in character format, the visit name, protocol-described of clinical encounter.
- ARMCD: Expected variable in character format, the planned arm code.
- ARM: Permissible variable in character format, the description of arm, name given to an Arm or Treatment Group.

Rule Variables:

- TVSTRL: Required variable in character format, the visit start rule.

TV Specification

Table 3.12: TV Specification

Variable	Label	Type	Length	Control Terminology	Origin	Notes
STUDYID	Study Identifier	char	20		Protocol: Set to "ABC-001"	
DOMAIN	Domain Abbreviation	char	2	DOMAIN	Assigned: Set to "TA"	
VISITNUM	Visit Number	char	20		Assigned	
VISIT	Visit Name	char	100		Protocol	
ARMCD	Planned Arm Code	char	100		Assigned	
ARM	Description of Planned Arm	char	100		Protocol	
TVSTRL	Visit Start Rule	char	100		Protocol	

SDTM.TV

Figure 3.32: SDTM.TV

	STUDYID	DOMAIN	VISITNUM	VISIT	ARM	ARMCD	TVSTRL
1	ABC-001	TV	1	SCREENING	DRUG A 10 mg	DRUG A	Inform Consent Signed
2	ABC-001	TV	2	DAY 1	DRUG A 10 mg	DRUG A	Admission to Clinic
3	ABC-001	TV	3	DAY 1	DRUG A 10 mg	DRUG A	Start of Day 1
4	ABC-001	TV	4	DAY 2	DRUG A 10 mg	DRUG A	Start of Day 2
5	ABC-001	TV	5	DAY 3	DRUG A 10 mg	DRUG A	Start of Day 3
6	ABC-001	TV	6	FOLLOW UP	DRUG A 10 mg	DRUG A	12 Days after last dose

3.7.5 Trial Inclusion/Exclusion Data Set (TI)

The Trial Inclusion/Exclusion (TI) domain describes the inclusion/exclusion criterion used to screen subjects. The structure of TI is one record per Inclusion/Exclusion criteria per trial.

Structure of Trial Inclusion/Exclusion (TI)

The structure of TI consists of Identifier, Topic, and Qualifier variables, as follows.

Identifier Variables:

- STUDYID: Required variable in character format, the unique study identifier. Each study has a unique study id. In the example study described in this book, we populate it with "ABC-001".
- DOMAIN: Required variable in character format, compliant with SDTM Implementation Guide DOMAIN sets to "TI".

Topic Variables:

- IETESTCD: Required variable in character format, the short name of IETEST. IETESTCD cannot contain characters other than letters, numbers, or underscores. The prefix "IE" is used to ensure consistency with IE domain.

Qualifier Variables:

- IETEST: Required variable in character format, the full text of inclusion/exclusion criteria.
- IECAT: Required variable in character format, the inclusion/exclusion category.

TI Specification

Table 3.13: TI Specification

Variable	Label	Type	Length	Control Terminology	Origin	Notes
STUDYID	Study Identifier	char	20		Protocol: Set to "ABC-001"	
DOMAIN	Domain Abbreviation	char	2	DOMAIN	Assigned: Set to "TA"	
IETESTCD	Inclusion/ Exclusion Criterion Short Name	char	20		Assigned	

(Continued)

Table 3.13: (*Continued*)

Variable	Label	Type	Length	Control Terminology	Origin	Notes
IETEST	Inclusion/ Exclusion Criterion	char	100		Protocol	
IECAT	Inclusion/ Exclusion Category	char	100		Assigned	

SDTM.TI

Figure 3.33: SDTM.TI

	STUDYID	DOMAIN	IETESTCD	IETEST	IECAT
1	ABC-001	TI	IN01	Subject will sign the inform consent form	INCLUSION
2	ABC-001	TI	IN02	Subjects are 18 years to 50 years of age	INCLUSION
3	ABC-001	TI	IN03	Female subjects are no child-bearing potential	INCLUSION
4	ABC-001	TI	EX01	Positive for HIV1 or HIV2 antibody during screening	EXCLUSION
5	ABC-001	TI	EX02	Cigarette smoking during the past 3 months at screening	EXCLUSION
6	ABC-001	TI	EX03	History of significant alcohol consumption in the past 3 months before screening	EXCLUSION

Chapter 4: Analysis Data Model (ADaM)

The Analysis Data Model (ADaM) is a standard structure, which is built from the SDTM. The fundamental principle of ADaM is data traceability, which means that the variables in ADaM track back from the corresponding SDTM domains, and from there back to the eCRF. Traceability establishes across data set and within data set relationships. For example, traceability facilitates the understanding of the relationship of the analysis variables from ADaM to its source data sets and variables in SDTM.

4.1 ADaM Standard Structures

The ADaM consists of three standard classes of data sets including Subject-Level Analysis Data Set (ADSL), Basic Data Structure (BDS), and Occurrence Data Structure (OCCDS). We will discuss these three types of ADaM data structures, the roles of variables that exist in ADaM data sets, and standard, common ADaM structure.

4.1.1 Variable "Roles"

Any variables in an ADaM data set whose name is the same as an SDTM variable must be a direct copy of the SDTM variable, and its label, meaning, and values must not be modified.

ADSL is an important domain that describes a subject's experience in the clinical trial, which is applied to all types of analysis. Some standard variables from SDTM.DM should be kept in ADSL, such as the reference start date (*RFSTDTC*), treatment arm (*ARM*), etc.

The BDS model is very flexible and allows almost any type of by visit data (LB, EG, VS, by-visit efficacy, etc.) to be created. In the BDS domain, ADaM data sets must have the variable parameter (*PARAM*) and analysis variable (*AVAL*). A subject can have one or more records, and the predecessors are the SDTM finding domains.

The OCCDS model is designed for occurrences that do not happen at a specific visit such as AE and CM. (This differs from the BDS structure mentioned above.) For the OCCDS structure, the predecessors are the SDTM Intervention class and Events class domains. Most of the variables in the OCCDS ADaM structures come directly from ADSL and corresponding SDTM occurrence domains including AE and CM.

4.1.2 Standard ADaM Structure and Domains

Some examples of ADaM structures and the domains within each structure are as follows. We go into more detail about each of these in subsequent sections.

Subject-level Analysis Data Set:

- ADSL.

Basic Data Structure Data Sets:

- ECG Test Results Analysis Data Set (ADEG); Laboratory Test Results Analysis Data Set (ADLB); Vital Sign Analysis Data Set (ADVS); Drug Accountability Analysis Data Set (ADDA).

Occurrence Data Structure Data Sets:

- Adverse Event Analysis Data Set (ADAE); Concomitant Medication Analysis Data Set (ADCM); Medical History Analysis Data Set (ADMH).

4.1.3 Standard ADaM Variables

- Required Variables: the variable must be included in the data set.
- Conditional Variables: the variable must be included in the data set in certain circumstances.
- Permissible Variable: the variable can be included in the data set depending on the specific needs from sponsors.

Note: All ADaM variables are populated as appropriate and null values are allowed, and we will describe when conditional and permissible variables need to be included in respective ADaM domains.

4.2 Subject-Level Analysis Data Set (ADSL)

ADSL contains one record per unique subject. It contains variables such as subject-level population flags such as a "Safety" flag for patients who were dosed, planned and actual treatment variables, demographic information, etc. ADSL combines required variables and other subject-level variables that describe a subject's experience in the clinical trial.

ADSL variables include variables from SDTM domains directly from DM as well as derived variables from EX, DS, and other SDTM domains.

4.2.1 Structure of ADSL

The structure of ADSL is one record per unique subject and consists of Identifier, Subject Demographics, Population flags, treatment, dose, and subject-level variables.

Identifier Variables:

- STUDYID: Required variable in character format, the predecessor is *DM.STUDYID*.
- USUBJID: Required variable in character format, the predecessor is *DM.USUBJID*.
- SUBJID: Required variable in character format, the predecessor is *DM.SUBJID*.
- SITEID: Required variable in character format, the predecessor is *DM.SITEID*.

ADSL Subject Demographics Variables:

- AGE: Required variable in character format, the predecessor is *DM.AGE*. If analysis needs require a derived age that does not match *DM.AGE,* then *AAGE* must be added.
- AGEU: Required variable in character format, the predecessor is *DM.AGEU*.
- SEX: Required variable in character format, the predecessor is *DM.SEX*.
- RACE: Required variable in character format, the predecessor is *DM.RACE*.

ADSL Population Flag Variables:

- FASFL: Conditional variable in character format, full analysis set population flag.
- SAFFL: Conditional variable in character format, safety population flag.

ADSL Treatment Variables:

- ARM: Required variable in character format, the predecessor is *DM.ARM.*
- TRTxxP: Required variable in character format, subject-level identifier that represents the planned treatment for period "xx".
- TRTxxA: Conditional variable in character format, subject-level identifier that represents the actual treatment for the subject for the period "xx".

ADSL Dose Variables:

- TRTSDT: Conditional variable in character format, date of first exposure to treatment for a subject in a study.
- TRTSDTM: Conditional variable in character format, datetime of first exposure to treatment for a subject in a study.

- TRTxxSDTF: Conditional variable in character format, date of the first exposure period "xx" input flag.
- TRTSDTF: Conditional variable in character format, date of the first exposure input flag.

ADSL Subject-Level Trial Experience Variables

- EOSDT: Permissible variable in character format, End of Study Date.
- EOSSTT: Permissible variable in character format, End of Study Status.
- Note: There are other groups of ADSL variables. We do not write about these variables in this book as they are not as commonly used and for the sake of brevity. For a more comprehensive understanding of these variables, please refer to CDISC's *Analysis Data Model Implementation Guide v 1.2 (ADaMIG 1.2).*

4.2.2 ADSL Specification

Table 4.1: ADSL Specification

Variable	Label	Type	Length	Controlled Terminology	Notes
STUDYID	Study Identifier	char	20		DM.STUDYID
USUBJID	Unique Subject Identifier	char	40		DM.USUBJID
SUBJID	Subject Identifier for the Study	char	20		DM.SUBJID
SITEID	Study Site Identifier	char	10		DM.SITEID
BRTHDTC	Date/Time of Birth	char	20		DM.BRTHDTC
AGE	Age	num	8		DM.AGE
AGEU	Age Units	char	10		DM.AGEU
SEX	Sex	char	2		DM.SEX
RACE	Race	char	100		DM.RACE

(Continued)

Table 4.1: (*Continued*)

Variable	Label	Type	Length	Controlled Terminology	Notes
ARACE	Analysis Race	char	100		If DM.RACE = "WHITE" then ARACE = "W"; else if DM.RACE = "BLACK OR AFRICAN AMERICAN" then ARACE = "B"; else if DM.RACE = "NATIVE HAWAIIAN OR OTHER PACIFIC ISLANDERS" then ARACE = "HP"; if DM.RACE = "ASIAN" then ARACE = "A"; else if DM.RACE = "AMERICAN INDIAN OR ALASKA AMERICAN" then ARACE = "AA"; else if race is not reported, then ARACE is null.
ETHNIC	Ethnicity	char	60		DM.ETHNIC
AETHNIC	Analysis Ethnicity	char	40		If DM.ETHNIC is null, then AETHNIC is "NOT COLLECTED", else is equal to DM.ETHNIC.
SAFFL	Safety Population Flag	char	1		If subjects received at least one dose then SAFFL = "Y", otherwise, set to "N".
FASFL	Full Analysis Set Population Flag	char	1		If SAFFL = "Y" then FASFL = "Y", otherwise, set to "N".
ARM	Description of Planned Arm	char	200		DM.ARM
ARMCD	Planned Arm Code	Char	20		DM.ARMCD
ACTARMCD	Actual Arm Code	char	20		DM.ACTARMCD
ACTARM	Description of Actual Arm	char	200		DM.ACTARM
TRT01P	Planned Treatment for Period 1	char	20		Set to TRTA, if ARMCD is "DRUG A".

(Continued)

Table 4.1: (*Continued*)

Variable	Label	Type	Length	Controlled Terminology	Notes
TRT01A	Actual Treatment for Period 1	char	20		Set to TRTA, if ACTARMCD is "DRUG A".
RFSTDTC	Subject Reference Start Date/Time	char	20		DM.RFSTDTC
RFENDTC	Subject Reference End Date/Time	char	20		DM.RFENDTC
TRTSDT	Date of First Exposure to Treatment	num	8		Numeric date part of DM.RFSTDTC
TRTSTM	Time of First Exposure to Treatment	num	8		Numeric time part of DM.RFSTDTC
TRT01SDTM	Date of First Exposure in Period 1	num	8		The earliest date of EX.EXSTDTC with numerical format
TRT01SDT	Datetime of First Exposure in Period 1	num	8		The date part of the TRT01SDTM
TRTEDT	Date of Last Exposure to Treatment	num	8		Numeric date part of DM.RFENDTC
TRTETM	Time of Last Exposure to Treatment	num	8		Numeric time part of DM.RFENDTC
TRT01EDTM	Date of Last Exposure in Period 1	num	8		The maximum date of EX.EXENDTC with numerical format
TRT01EDT	Datetime of Last Exposure in Period 1	num	8		

(Continued)

Table 4.1: (*Continued*)

Variable	Label	Type	Length	Controlled Terminology	Notes
EOSSTT	End of Study Status	char	20		The subject's status at the end of the study or data cutoff. Set to "COMPLETED" when DS.DSDECOD= "COMPLETED". Set to "DISCONTINUED" when DS.DSDECOD is not equal to "COMPLETED".
EOSDT	End of Study Date	num	8		Date of study ends. Equal to DS.DSSTDTC in numeric format.
COUNTRY	Country	char	4	COUNTRY	DM.COUNTRY

4.2.3 ADSL Programming

In order to create the ADSL domain, we start with SDTM.DM. Please see code below.

Note: For every domain, the corresponding SAS program name is domain.sas. For example, ADSL is created by adsl.sas, so all code in this section belongs to adsl.sas.

```
/*Begin writing SAS program sdtm.dm.sas*/
/*Demographic Variables*/
data dm1;
  set sdtm.dm; /*Derive AETHNIC, ARACE */

  length AETHNIC $40.;
  if missing(ETHNIC) then AETHNIC = "NOT COLLECTED";
  else AETHNIC = ETHNIC;
  If RACE = "WHITE" then ARACE = "W";
  else if RACE = "BLACK OR AFRICAN AMERICAN" then ARACE = "B";
  else if RACE = "NATIVE HAWAIIAN OR OTHER PACIFIC ISLANDERS" then ARACE = "HP";
  else if RACE = "ASIAN" then ARACE = "A"; else if RACE = "AMERICAN INDIAN OR
ALASKA AMERICAN" then ARACE = "AA"; ❶

/*Derive TRTSDT, TRTSTM, TRTEDT, TRTETM */
  TRTSDT = input(substr(RFSTDTC,1,10),yymmdd10.);TRTSTM =
input(substr(RFSTDTC,12),time5.);
  TRTEDT = input(substr(RFENDTC,1,10),yymmdd10.);
  TRTETM = input(substr(RFENDTC,12),time5.);
  format TRTSDT yymmdd10. TRTSTM time5. TRTEDT yymmdd10. TRTETM time5.; ❷
```

```
   length TRT01P TRT01A $20.;

   /*Derive TRT01P, TRT01A*/
   if ARMCD = "DRUG A" then TRT01P = "TRTA";
   if ACTARMCD = "DRUG A" then TRT01A = "TRTA"; /*Derive SAFFL, FASFL */

   if ^missing(RFSTDTC) then do;
      SAFFL="Y";
      FASFL="Y";
   end;
   else do;
      SAFFL="N";
      FASFL="N";
   end;
run;
```

❶ Based on the variables in SDTM.DM, we derive variables AETHNIC, ARACE, and variables related to reference treatment date, which is usually treatment start date, and time including TRTSDT, TRTSTM, TRTEDT, TRTETM, TRT01P, TRT01A.

Note: *TRTSDT* is derived from the SDTM domain *DM.RFSTDTC*, which is typically the first date in the EX domain, whereas *TRT01P* is from the SDTM TA domain, which is the planned treatment, such as "Drug A".

❷ For the following numeric date and time variables (*TRTSDT* and *TRTSTM*, *TRTEDT* and *TRTETM,*) we use the input() function to convert character to numeric format first, then we use the SAS FORMAT statement to specify the format of the new variables *TRTSDT, TRTSTM, TRTEDT, TRTETM*. Lastly, we derive two population flags *SAFFL* and *FASFL*.

```
/*Merge dm2 and dm1 to catch DSDECOD, DSSTDTC*/
proc sql;
   create table dm2 as select a.*,b.DSDECOD,DSSTDTC from dm1 a left join
sdtm.ds b on a.USUBJID=b.USUBJID;                              ❶
quit;
/*Derive EOSDT, EOSSTT variables*/

data dm3;
   set dm2;
   length EOSDT 8 EOSSTT $20.;
   EOSDT = input(substr(DSSTDTC,1,10),yymmdd10.);
   if DSDECOD = "COMPLETED" then EOSSTT = "COMPLETED";
   else EOSSTT = "DISCONTINUED";
   format EOSDT yymmdd10.;
run;
/*Sort data set SDTM.EX by USUBJID, EXSTDTC without duplicate values*/

proc sort data=SDTM.EX out=ex nodupkey;
   by USUBJID EXSTDTC EXENDTC;
run;

data ex1
     ex2;
```

```
  set ex;
  by USUBJID EXSTDTC EXENDTC;

  /*Create data set ex1 if the first.USUBJID statement*/
  /*Create data set ex2 if the last.USUBJID statement*/
  if First.USUBJID then output ex1;
  if Last.USUBJID then output ex2;
run;

/*Create data set dm4 and dm5*/
proc sql;                     ❷
  create table dm4 as select a.*, b.EXSTDTC from dm3 a left join ex1 b on
a.USUBJID=b.USUBJID;
  create table dm5 as select a.*, b.EXENDTC from dm4 a left join ex2 b on
a.USUBJID=b.USUBJID;
quit;

data Final;
  set dm5;
  /*Derive TRT01SDTM, TRT01SDT, TRT01EDTM, TRT01EDT */
  length TRT01SDTM TRT01SDT TRT01EDTM TRT01EDT 8.;     ❸
  TRT01SDTM = input(EXSTDTC, e8601dt.);
  TRT01SDT = input(substr(EXSTDTC,1,10),yymmdd10.);
  TRT01EDTM = input(EXENDTC, e8601dt.);
  TRT01EDT = input(substr(EXENDTC,1,10),yymmdd10.);
  format TRT01SDTM TRT01EDTM e8601dt. TRT01SDT TRT01EDT yymmdd10.;
run;
```

❶ Merge DM1 and DS data sets to capture *DSDECOD* and *DSSTDTC* variables, then derive *EOSDT* and *EOSSTT* based on the ADSL specification. (Please see Table 4.1.)

❷ Merge DM3 and EX1 data sets to capture *EXSTDTC* variable, then use if SAS First. statement logic to get the minimum value of *EXSTDTC* using *USUBJID* and *EXSTDTC*. Merge DM4 and EX2 data sets to include the *EXENDTC* variable, and as with the SAS First. statement logic example above, use the if Last. Statement logic to get the maximum value of *EXENDTC* using *USUBJID* and *EXENDTC* variables.

❸ Derive *TRT01SDTM*, *TRT01SDT*, *TRT01EDTM,* and *TRT01EDT* variables.

```
libname ADAM "...directory";
data ADAM.ADSL(label="Subject-Level Analysis Data Set");

/*Assign variable attributes such as label and length to conform with ADAM.
ADSL Specification (these will also be the same attributes as the ADAM IG).*/

attrib
    STUDYID     label = "Study Identifier"                     length = $20
    USUBJID     label = "Unique Subject Identifier"            length = $40
    SUBJID      label = "Subject Identifier for the Study"     length = $20
    SITEID      label = "Study Site Identifier"                length = $10
    BRTHDTC     label = "Date/Time of Brith"                   length = $20
    AGE         label = "Age"                                  length = 8
    AGEU        label = "Age Units"                            length = $10
```

```
   SEX          label = "Sex"                                    length = $2
   RACE         label = "Race"                                   length = $100
   ARACE        label = "Analysis Race"                          length = $100
   ETHNIC       label = "Ethnicity"                              length = $60
   AETHNIC      label = "Analysis Ethnicity"                     length = $60
   SAFFL        label = "Safety Population Flag"                  length = $1
   FASFL        label = "Full Analysis Set Population Flag"       length = $1
   ARM          label = "Description of Planned Arm"             length = $200
   ARMCD        label = "Planned Arm Code"                        length = $20
   ACTARMCD     label = "Actual Arm Code"                         length = $20
   ACTARM       label = "Description of Actual Arm"              length = $200
   TRT01P       label = "Planned Treatment for Period 1"          length = $20
   TRT01A       label = "Actual Treatment for Period 1"           length = $20
   RFSTDTC      label = "Subject Reference Start Date/Time"       length = $20
   RFENDTC      label = "Subject Reference End Date/Time"         length = $20
   TRTSDT       label = "Date of First Exposure to Treatment"     length = 8
   TRTSTM       label = "Time of First Exposure to Treatment"     length = 8
   TRT01SDTM    label = "Datetime of First Exposure in Period 1"  length = 8
   TRT01SDT     label = "Date of First Exposure in Period 1"      length = 8
   TRTEDT       label = "Date of Last Exposure to Treatment"      length = 8
   TRTETM       label = "Time of Last Exposure to Treatment"      length = 8
   TRT01EDTM    label = "Datetime of Last Exposure in Period 1"   length = 8
   TRT01EDT     label = "Date of Last Exposure in Period 1"       length = 8
   EOSSTT       label = "End of Study Status"                     length = $20
   EOSDT        label = "End of Study"                            length = 8
   COUNTRY      label = "Country"                                 length = $4
   ;
 set Final;
  keep STUDYID USUBJID SUBJID SITEID BRTHDTC AGE AGEU SEX RACE ARACE ETHNIC
AETHNIC SAFFL FASFL ARM ARMCD ACTARMCD ACTARM TRT01P TRT01A RFSTDTC RFENDTC
TRTSDT TRTSTM TRT01SDTM TRT01SDT TRTEDT TRTETM TRT01EDTM TRT01EDT EOSSTT EOSDT
COUNTRY
     ;
run;
```

Finally, assign the label and length to each variable to conform with ADAM.ADSL Specification. (These will also be the same attributes as the ADaM IG.)

Note: For the purposes of this book, we refer to SAS LIBNAME in uppercase; however, in your day-to-day programming work, we recommend using lowercase for both LIBNAME statements and data sets.

The final ADSL data set is saved in the permanent SAS library, called ADAM, as shown in Figure 4.1.

Figure 4.1: ADAM.ADSL

	STUDYID	USUBJID	SUBJID	SITEID	BRTHDTC
1	ABC-001	ABC-001-001-001	001	001	2000-02-03
2	ABC-001	ABC-001-001-002	002	001	2000-04-06
3	ABC-001	ABC-001-001-003	003	001	1990-08-11
4	ABC-001	ABC-001-001-004	004	001	1985-09-09
5	ABC-001	ABC-001-001-005	005	001	1998-02-03
6	ABC-001	ABC-001-001-006	006	001	1986-06-09
7	ABC-001	ABC-001-001-007	007	001	1996-02-01
8	ABC-001	ABC-001-001-008	008	001	2000-09-05
9	ABC-001	ABC-001-001-009	009	001	1990-07-08
10	ABC-001	ABC-001-001-010	010	001	1990-08-20

AGE	AGEU	SEX	RACE	ARACE	ETHNIC	AETHNIC
20	YEARS	F	WHITE	W	NOT HISPANIC OR LATINO	NOT HISPANIC OR LATINO
20	YEARS	F	WHITE	W	NOT HISPANIC OR LATINO	NOT HISPANIC OR LATINO
30	YEARS	M	WHITE	W	NOT HISPANIC OR LATINO	NOT HISPANIC OR LATINO
35	YEARS	M	WHITE	W	NOT HISPANIC OR LATINO	NOT HISPANIC OR LATINO
22	YEARS	M	WHITE	W	NOT HISPANIC OR LATINO	NOT HISPANIC OR LATINO
34	YEARS	M	WHITE	W	NOT HISPANIC OR LATINO	NOT HISPANIC OR LATINO
24	YEARS	M	WHITE	W	NOT HISPANIC OR LATINO	NOT HISPANIC OR LATINO
20	YEARS	M	BLACK OR AFRICAN AMERICAN	B	NOT HISPANIC OR LATINO	NOT HISPANIC OR LATINO
30	YEARS	M	BLACK OR AFRICAN AMERICAN	B	NOT HISPANIC OR LATINO	NOT HISPANIC OR LATINO
30	YEARS	M	BLACK OR AFRICAN AMERICAN	B	NOT HISPANIC OR LATINO	NOT HISPANIC OR LATINO

SAFFL	FASFL	ARM	ARMCD	ACTARMCD	ACTARM	TRT01P	TRT01A
Y	Y	DRUG A 10 MG	DRUG A	DRUG A	DRUG A 10 MG	TRTA	TRTA
Y	Y	DRUG A 10 MG	DRUG A	DRUG A	DRUG A 10 MG	TRTA	TRTA
Y	Y	DRUG A 10 MG	DRUG A	DRUG A	DRUG A 10 MG	TRTA	TRTA
Y	Y	DRUG A 10 MG	DRUG A	DRUG A	DRUG A 10 MG	TRTA	TRTA
Y	Y	DRUG A 10 MG	DRUG A	DRUG A	DRUG A 10 MG	TRTA	TRTA
Y	Y	DRUG A 10 MG	DRUG A	DRUG A	DRUG A 10 MG	TRTA	TRTA
Y	Y	DRUG A 10 MG	DRUG A	DRUG A	DRUG A 10 MG	TRTA	TRTA
Y	Y	DRUG A 10 MG	DRUG A	DRUG A	DRUG A 10 MG	TRTA	TRTA
Y	Y	DRUG A 10 MG	DRUG A	DRUG A	DRUG A 10 MG	TRTA	TRTA
Y	Y	DRUG A 10 MG	DRUG A	DRUG A	DRUG A 10 MG	TRTA	TRTA

RFSTDTC	RFENDTC	TRTSDT	TRTSTM	TRT01SDTM	TRT01SDT
2021-01-02T10:00:00	2021-01-04T10:00:00	2021-01-02	10:00	2021-01-02T10:00:00	2021-01-02
2021-01-02T10:00:00	2021-01-04T10:00:00	2021-01-02	10:00	2021-01-02T10:00:00	2021-01-02
2021-01-02T10:00:00	2021-01-04T10:00:00	2021-01-02	10:00	2021-01-02T10:00:00	2021-01-02
2021-01-02T10:00:00	2021-01-04T10:00:00	2021-01-02	10:00	2021-01-02T10:00:00	2021-01-02
2021-01-02T10:00:00	2021-01-04T10:00:00	2021-01-02	10:00	2021-01-02T10:00:00	2021-01-02
2021-01-02T10:00:00	2021-01-04T10:00:00	2021-01-02	10:00	2021-01-02T10:00:00	2021-01-02
2021-01-02T10:00:00	2021-01-04T10:00:00	2021-01-02	10:00	2021-01-02T10:00:00	2021-01-02
2021-01-02T10:00:00	2021-01-04T10:00:00	2021-01-02	10:00	2021-01-02T10:00:00	2021-01-02
2021-01-02T10:00:00	2021-01-04T10:00:00	2021-01-02	10:00	2021-01-02T10:00:00	2021-01-02
2021-01-02T10:00:00	2021-01-04T10:00:00	2021-01-02	10:00	2021-01-02T10:00:00	2021-01-02

TRTEDT	TRTETM	TRT01EDTM	TRT01EDT	EOSSTT	EOSDT	COUNTRY
2021-01-04	10:00	2021-01-04T10:00:00	2021-01-04	COMPLETED	2021-01-04	USA
2021-01-04	10:00	2021-01-04T10:00:00	2021-01-04	COMPLETED	2021-01-04	USA
2021-01-04	10:00	2021-01-04T10:00:00	2021-01-04	COMPLETED	2021-01-04	USA
2021-01-04	10:00	2021-01-04T10:00:00	2021-01-04	COMPLETED	2021-01-04	USA
2021-01-04	10:00	2021-01-04T10:00:00	2021-01-04	COMPLETED	2021-01-04	USA
2021-01-04	10:00	2021-01-04T10:00:00	2021-01-04	COMPLETED	2021-01-04	USA
2021-01-04	10:00	2021-01-04T10:00:00	2021-01-04	COMPLETED	2021-01-04	USA
2021-01-04	10:00	2021-01-04T10:00:00	2021-01-04	COMPLETED	2021-01-04	USA
2021-01-04	10:00	2021-01-04T10:00:00	2021-01-04	COMPLETED	2021-01-04	USA
2021-01-04	10:00	2021-01-04T10:00:00	2021-01-04	COMPLETED	2021-01-04	USA

4.3 Basic Data Structure (BDS)

The structure of BDS contains one or more records per subject, per analysis parameter, per analysis time point. Parameter is a term used in ADaM, which describes the analysis parameter. It must include all descriptive and qualifying information relevant to the analysis purpose of the parameter.

The BDS is designed to facilitate the majority of analysis required for any type of study – from lab and vital sign results to all types of efficacy endpoints. The only ADaM data that is not created in BDS is ADSL and "occurrence" data, such as any events that are not scheduled at a planned visit (e.g., ADAE and AECM).

All finding class data sets are created using the BDS design. In addition to the essential variables such as *STUDYID, DOMAIN*, and *USUBJID,* there are the following additional categories of variables, including: *XXTESTCD, XXTEST, XXORRES, XXSTRESN, XXSTRESC, XXDTC, XXBLFL, VISITNUM* and *VISIT* across all BDS domains. ("XX" represents the primary SDTM domain used to create the corresponding ADaM domain.)

Further, there are other common variables in ADaM BDS domains such as *PARAMCD, PARAM, AVAL, AVALC, ADT, ABLFL, AVISITN*, and *AVIST.* In particular, *XXTESTCD* and *XXTEST* are used to generate *PARAMCD* and *PARAM*, respectively; *XXSTRESN* is used to generate *AVAL; XXBLFL* is used to generate *ABLFL; VISIT* and *VISITNUM* are used to generate *AVISIT* and *AVISITN*. This is demonstrated later in the chapter in the SAS code sections.

In addition, there are variables in ADaM that need to be copied directly from the ADSL data set, such as *SITEID, COUNTRY, BRTHDTC, AGE, SEX, RACE, ETHNIC,* Population flags and treatment-related information (planned and actual treatment, date of first treatment, etc.). For a more comprehensive understanding of these variables, please refer to the *ADaM IG 1.2.*

4.3.1 ECG Test Results Analysis Data Sets (ADEG)

Structure of ADEG

As mentioned earlier, the structure of ADEG is BDS and therefore contains one or more records per subject, per analysis parameter, per analysis time point.

Identifier Variables:

- STUDYID: Required variable in character format, the predecessor is *EG.STUDYID.*
- USUBJID: Required variable in character format, the predecessor is *EG.USUBJID.*
- ASEQ: Permissible variable in character format, Analysis Sequence Number, to ensure uniqueness of subject records within an ADaM data set.

Traceability Variables:

- EGSEQ: Permissible variable in character format, the predecessor is *EG.EGSEQ*.

Record-Level Treatment Variables:

- TRTP: Conditional variable in character format, planned treatment, derived from ADSL.
- TRTA: Conditional variable in character format, actual treatment, derived from ADSL.

Timing Variables:

- ADT: Permissible variable in character format, analysis date.
- ATM: Permissible variable in character format, analysis time.
- ADTM: Permissible variable in character format, analysis datetime.
- ADY: Permissible variable in character format, analysis relative day of *AVAL* and/or *AVALC*.
- ASTDT: Permissible variable in character format, analysis start date associated with *AVAL* and/or *AVALC*.
- AVISIT: Conditional variable in character format, the analysis visit description, required if an analysis is done by normal, assigned, or analysis visit.
- AVISITN: Conditional variable in numeric format, representative of *AVISIT*.
- APHASE: Conditional variable in character format, is a record-level timing variable that represents the analysis period within the study associated with the record for analysis purposes.
- VISIT: Required variable in character format, the predecessor is *LB.VISIT*.
- VISITNUM: Required variable in numeric format, the predecessor is *LB.VISITNUM*.

Analysis Parameter Variables:

- PARAM: Required variable in character format, the description of the analysis parameter. It must include all descriptive and qualifying information relevant to the analysis purpose of the parameter.
- PARAMCD: Required variable in character format, the short name of analysis parameter in *PARAM*.
- AVAL: Conditional variable in numeric format, analysis value, analysis value described by *PARAM*.
- AVALC: Conditional variable in character format, character analysis value described by *PARAM*, *AVALC* can be a character string mapping to *AVAL*.
- BASE: Conditional variable in character format, baseline value. The subject's baseline analysis value for a parameter and baseline definition (BASETYPE) if presented. BASE contains the value of *AVAL* copied from a record within the parameter on which *ABLFL*= "Y".
- CHG: Permissible variable in numeric format, which is the change from the baseline analysis value. It is equal to *AVAL- BASE*. If used for a given *PARAM, CHG* should be populated for all post-baseline records of that *PARAM* regardless of whether that record is used for analysis.

Analysis Descriptor Variables for BDS Data Sets:

- DTYPE: Conditional variable in character format, which is the derivation type, *and* is used to denote, and must be populated when the value of *AVAL* or *AVALC* has been imputed or derived differently than the other analysis values within the parameter. DTYPE is required to be populated even if *AVAL* and *AVALC* are null for any derived record.

Indicator Variables for BDS Data Sets:

- ABLFL: Conditional variable in character format, which is a character indicator to identify the baseline record for the subject, parameter, and baseline type combination. *ABLFL* is required if *BASE* is present in the data set.
- Note: There are other types of variables in the ADaM BDS data sets, such as "Analysis Parameter Criterion Variables", "Toxicity and Range Variables for BDS Data sets", and "BDS population indicator". For a more comprehensive understanding of these variables, please refer to the *ADaM IG 1.2*.

ADEG Specification

Table 4.2: ADEG Specification

Variable	Label	Type	Length	Controlled Terminology	Notes
STUDYID	Study Identifier	char	20		EG.STUDYID
USUBJID	Unique Subject Identifier	char	40		EG.USUBJID
EGSEQ	Sequence Number	num	8		EG.ESEQ
ASEQ	Analysis Sequence Number	num	8		Sort by USUBJID, PARAMCD, ADT, ATM, DTYPE, EGSEQ, start with 1 for the first value for the first row for each subject and increment by 1 at each successive record.
TRTP	Planned Treatment	char	40		Equal to ADSL.TRT01P
TRTA	Actual Treatment	char	40		Equal to ADSL.TRT01A
ADT	Analysis Date	num	8		Equal to numeric date part of EG.EGDTC

(Continued)

Table 4.2: (*Continued*)

Variable	Label	Type	Length	Controlled Terminology	Notes
ATM	Analysis Time	num	8		
ADTM	Analysis Date and Time	num	8		Equal to EG.EGDTC
ADY	Analysis Relative Day	num	8		Equal to the difference between ADT and ADSL.TRTSDT, if ADT is on or after ADSL.TRTSDT, then add 1.
AVISIT	Analysis Visit	char	40		Equal to "Baseline" if ABLFL is "Y"; Equal to "Follow-Up" if VISIT is "Follow Up"; else equal to VISIT.
AVISITN	Analysis Visit (N)	num	8		Equal to 0, if AVISIT is "Baseline"; equal to 100, if AVISIT is "Follow-Up"; else equal to X (depending on which visit it is and what the sponsor convention is).
APHASE	Phase	char	40		Equal to "Screening" when the assessment is on or before the first dose of study drug; equal to "Treatment" when the assessment is after the first dose of study drug. Equal to "Follow-Up" when the assessment is after the last dose date.
PARAM	Parameter	char	40		Concatenate EG.EGTEST with EG.EGORRESU
PARAMCD	Parameter Code	char	8		EG.EGTESTCD

(*Continued*)

Table 4.2: (*Continued*)

Variable	Label	Type	Length	Controlled Terminology	Notes
AVAL	Analysis Value	num	8		When PARAMCD is not equal to "INTP" and DTYPE is equal to "AVERAGE", then equal to the average of multiple assessments. When PARAMCD not is equal to "INTP" and DTYPE is null, then AVAL is equal to EG.EGSTRESC with numeric format.
AVALC	Analysis Value (C)	char	40		When PARAMCD is not equal to "INTP", then AVALC is null; when PARAMCD is equal to "INTP", then AVALC is equal to "ABNORMAL" if EG.EGSTRESC is "ABNORMAL; AVAL is equal to NORMAL if EG.EGSTRESC is "NORMAL".
ABLFL	Baseline Record Flag	char	1		Only populate when DTYPE is "AVERAGE", equal to "Y" for the most recent nonmissing measurement before the first dose of study drug.
BASE	Baseline Value	num	8		Only populate baseline and post-baseline (EMDESC in ("P","T")) when ABLFL is " Y", equal to AVAL, not populate when EMDESC is "A".

(*Continued*)

Table 4.2: (*Continued*)

Variable	Label	Type	Length	Controlled Terminology	Notes
CHG	Change from Baseline	num	8		Only populate post-baseline (EMDESC in "T") when AVAL and BASE are nonmissing, equal to (AVAL - BASE).
DTYPE	Derivation Type	char	20		Equal to "AVERAGE" when AVAL is derived value.
EMDESC	Description of Treatment Emergent	char	20		Equal to "P" when the assessment is on or before the first dose of study drug (ADT<=ADSL.TRTSDT and ATM is missing. If ADT and ATM is not missing, then ADTM<=ADSL.TRT01SDTM); equal to "T" when the assessment is after the first dose of study drug (ADT > ADSL.TRTSDT and ATM is missing, or if ADT and ATM is not missing then ADTM>ADSL.TRT01SDTM); otherwise, equal to "A" when the assessment is after the end of the treatment
EGORRES	Result of Finding in Original Units	char	100		EG.EGORRES
EGORRESU	Original Units	char	40		EG.EGORRESU
EGSTRESC	Character Result/ Finding in Std Format	char	100		EG.EGSTRESC

(*Continued*)

Table 4.2: (*Continued*)

Variable	Label	Type	Length	Controlled Terminology	Notes
VISIT	Visit Name	char	40		EG.VISIT
VISITNUM	Visit Number	num	8		EG.VISITNUM
EGDTC	Date/Time of ECG	char	40		EG.EGDTC; equal to null for the average records

ADEG Programming

In order to create the ADEG domain, we start with by merging SDTM.EG with ADAM.ADSL. Please see code below.

Note: For every domain, the corresponding SAS name is domain.sas. For example, ADEG is created by adeg.sas, so all code in this section belongs to adeg.sas.

```
/*Begin writing SAS program merge SDTM.EG and ADAM.ADSL.sas*/

data ADEG1;
  merge SDTM.EG(in=a) ADAM.ADSL(in=b drop=STUDYID);    ❶
  by USUBJID;
  if a and b;
run;

proc sql;
   create table ADEG2 as select distinct EGTEST,EGORRESU from ADEG1 where
EGORRESU ne '';
quit;

data ADEG3;
  length PARAM $40;         ❷
  set ADEG2;
  PARAM=strip(EGTEST)||" ("||strip(EGORRESU)||")";
  keep EGTEST PARAM;
run;

proc sql;          ❸
/*Left Join PARAM from ADEG1 with ADEG3 when the same EGTEST*/
create table ADEG4 as select a.*,b.PARAM from ADEG1 as a left join ADEG3 as b
on a.EGTEST=b.EGTEST;

/*Map USUBJID, EGTESTCD, and the number of USUBJID from ADEG4 to ADEG5*/

create table ADEG5 as select USUBJID,EGTESTCD,count(USUBJID) as count from ADEG4
group by USUBJID,EGTESTCD;

/*Left Join ADEG4 with ADEG5 when the USUBJID and EGTESTCD are the same */
```

```
create table ADEG6 as select a.*,b.count from ADEG4 as a left join ADEG5 as b
on a.USUBJID=b.USUBJID and a.EGTESTCD=b.EGTESTCD;
quit;
```

❶ Merge SDTM.EG with ADAM.ADSL since there are many variables from ADSL when ADEG is derived.
❷ Capture *EGTEST* and *EGORRESU* to create *PARAM*.
❸ Merge *PARAM* from ADEG3 data set back to ADEG1 data set, then use the SAS count function by *USUBJID* and *EGTESTCD*. Refer to ADAM.ADEG Specification. Finally, merge by the *COUNT* variable from ADEG4 back to ADEG5 to get ADEG6 data set.

```
/*Derive DTYPE*/
data ADEG7;
  set ADEG6(where=(EGTESTCD^="INTP"));
  length DTYPE $20;
  EGSTRESN=input(EGSTRESC,best.);
  by USUBJID EGTESTCD;
  if first.EGTESTCD then do;
    SUM=EGSTRESN;
    N=1;
  end;
  else do ;
    SUM+EGSTRESN;
    N+1;
  end;
  output;
  if last.EGTESTCD then do; DTYPE="AVERAGE"; FLAG=1;output; end;
run;
```

In order to help the reader understand the internal process SAS uses when deriving *EGSTRSTN* and *DTYPE* variables, ADEG7 output is shown in Figure 4.2.

Figure 4.2: ADEG7 Output

```
/*Derive EGSTRESN and EGSTRESC*/


data ADEG8;
  set ADEG7;
```

```
   if DTYPE="AVERAGE" then do;
   if SUM ne . then do;
     EGSTRESN_MEAN=SUM/N;                    ❶
     EGSTRESC=strip(put(EGSTRESN,best.));
   end;
   else  do;
     EGSTRESN=.;
     EGSTRESC="";
   end;
end;
end;
run;


/*Set ADEG8 and ADEG6 with EGTESTCD="INTP"*/


data ADEG9;
  set ADEG8 ADEG6(where=(EGTESTCD="INTP"));
run;
```

❶ Derive the average value for multiple assessments: *EGSTRESN_MEAN* and *EGSTRESC*. Then merge ADEG8 when *EGTESTCD* is not equal to "INTP" and ADEG6 when *EGTESTCD* is equal to "INTP".

Based on the variables in ADAM.ADEG, the following code derives some analysis variables including *ADT, ATM, ADTM, ADY, AVAL, AVALC, TRTP, TRTA, APHASE*, and *EMDESC*.

```
/*Sort data set ADEG9 by USUBJID, EGTESTCD, EGDTC */
proc sort data=ADEG9;
      by USUBJID EGTESTCD EGDTC;
run;

data ADEG10;
  set ADEG9;

  /*Derive ADT, ATM, ADTM */
  length PARAMCD $8. AVALC $40.;
  if length(EGDTC)=10 then do;
     ADT=input(EGDTC,yymmdd10.);ATM=.;ADTM=.;
end;

  if length(EGDTC)>10 then do;
     ADTM=input(EGDTC,is8601dt.);
     ADT=datepart(ADTM);
     ATM=timepart(ADTM);
  end;
  format ADTM is8601dt. ADT yymmdd10. ATM time5.;
  /*Derive ADY */
  if nmiss(ADT,TRTSDT)=0 then ADY=ADT-TRTSDT+(ADT>=TRTSDT);
  /*Derive APHASE,EMDESC*/

  if (ADT<=TRTSDT and ATM=.) or (ADT^=. and ATM^=. and ADTM<=TRT01SDTM) then do;
     APHASE="Screening";
```

```
        EMDESC="P";
    end;
    if (ADT > TRTSDT and ATM =.) or (ADT^=. and ATM^=. and ADTM>TRT01SDTM) then
do;
        APHASE="Treatment";
        EMDESC="T";
    end;
    if (ADT > TRTEDT and ATM =.) or (ADT^=. and ATM^=. and ADTM>TRT01EDTM) then
do ;
      APHASE="Follow-Up";
      EMDESC="A";
    end;

    /*Derive PARAMCD */
    if EGTEST='Interpretation' then PARAM=strip(EGTEST);
    else PARAM=PARAM;
    PARAMCD=strip(EGTESTCD);
    /*Derive AVAL and AVALC */

    if PARAMCD^='INTP' and DTYPE='AVERAGE' then AVAL=EGSTRESN_MEAN;
    else if PARAMCD^='INTP' and DTYPE='' then AVAL=EGSTRESN;
    else if PARAMCD='INTP' then AVAL=.;
    if PARAMCD='INTP' then AVALC=strip(EGSTRESC);
    else if PARAMCD^='INTP' then AVALC='';
    /*Derive TRTP, TRTA */

    TRTP=TRT01P;
    TRTA=TRT01A;

run;
```

The following code is to count *NUMBER* for the purpose of deriving *ABLFL* and, later, merging *ABLFL* variable back to ADEG10 based on the same *NUMBER*.

```
/*Derive ABLFL */
data ADEG10;
  set ADEG10;
NUMBER=_n_;
run;

/*Sort data set ADEG10 by USUBJID, PARAMCD, ADT, ADTM, DTYPE */
proc sort data=ADEG10;
  by USUBJID PARAMCD ADT ADTM DTYPE;
run;

/*Filter the condition of baseline flag */

data BASE;
  set ADEG10(where=(EMDESC="P" and (AVAL ne . or AVALC ne '')  and
(.<ADT<=TRTSDT) and COUNT>1));
  by USUBJID PARAMCD ADT ADTM DTYPE;              ❶
```

```
run;

 /*if the last PARAMCD then ABLFL sets to "Y" */

data ABLFL;
  set BASE;
  by USUBJID PARAMCD ADT ADTM DTYPE;
  if last.PARAMCD; ABLFL="Y";
run;
/*Left Join ADEG10 with ABLFL */

proc sql;                    ❷
      create table ADEG11 as select a.*,b.ABLFL from ADEG10 as a left join
ABLFL as b on a.NUMBER=b.NUMBER;
quit;
```

❶   Refer to ADAM.ADEG specification. *ABLFL* is equal to "Y" for the most recent
    nonmissing measurement before the first dose of study drug.

❷   Merge *ABLFL* from ADEG10 back to ADEG11 based on the same *NUMBER*.

```
/*Derive variable Base */
proc sql;
      create table ADEG12 as select a.*,b.AVAL as BASE from ADEG11 as a
    left join ADEG11(where=(ABLFL='Y')) as b on a.USUBJID=b.USUBJID and
a.PARAMCD=b.PARAMCD;
quit;              ❸
```

❸   The variable *BASE* is equal to *AVAL* when last nonmissing predose (ABLFL= "Y").

```
 /*Derive variable CHG*/

data ADEG13;
  set ADEG12;
  if n(AVAL,BASE)=2 and ABLFL ne "Y"  then CHG=AVAL-BASE;     ❹
  if ABLFL^="Y" and EMDESC="P" then do; CHG=.;end;
run;
```

❹   Derive *CHG* (change from baseline), which is only populated for the post-baseline records.

```
/*Derive variable AVISIT and AVISITN*/

data ADEG14;
  set ADEG13;
  length AVISIT $40. AVISITN 8. ;
  if PARAMCD ne "INTP" then do;
  if ABLFL="Y" then do;
    AVISIT="Baseline";
    AVISITN=0;
  end;
  else if index(VISIT,"FOLLOW-UP") then do;
    AVISIT="Follow-up";
    AVISITN=100;
  end;
```

```
   else do;
     AVISIT=strip(VISIT);
     AVISITN=input(compress(AVISIT,,"kd"),best.);   ❺
   end;
 end;
run;
```

❺ Derive *AVISIT, AVISITN*; we use the SAS function compress () with "kd" to only keep numeric values for *AVISIT*.

The following code creates the *ASEQ* variable. This variable is used to identify the unique analysis sequence number of subjects within a domain.

```
/*Sort data set ADEG14*/

proc sort data=ADEG14;
 by USUBJID PARAMCD ADT ATM DTYPE EGSEQ;
run;

/*Derive ASEQ*/
data Final;
  set ADEG14;
  by USUBJID PARAMCD ADT ATM DTYPE EGSEQ;
  if first.USUBJID then ASEQ = 0;
    ASEQ+1;
  output;
run;
```

Create variable *ASEQ* variable. This variable is used to uniquely record analysis sequence number of subjects within a domain.

```
libname ADAM ".../directory";
data ADAM.ADEG(label="ECG Test Results Analysis Data sets");

/*Assign variable attributes such as label and length to conform with ADAM.
ADSL Specification (these will also be the same attributes as the ADAM IG).*/

attrib
    STUDYID     label = "Study Identifier"                      length = $20
    USUBJID     label = "Unique Subject Identifier"             length = $40
    SUBJID      label = "Subject Identifier for the Study"      length = $20
    EGSEQ       label = "Sequence Number"                       length = 8
    ASEQ        label = "Analysis Sequence Number"              length = 8
    TRTP        label = "Planned Treatment"                     length = $40
    TRTA        label = "Actual Treatment"                      length = $40
    ADT         label = "Analysis Date"                         length = 8
    ATM         label = "Analysis Time"                         length = 8
    ADTM        label = "Analysis Date and Time"                length = 8
    ADY         label = "Analysis Relative Day"                 length = 8
    AVISIT      label = "Analysis Visit"                        length = $40
    AVISITN     label = "Analysis Visit (N)"                    length = 8
    APHASE      label = "PHASE"                                 length = $40
```

```
     PARAM       label = "Parameter"                               length = $40
     PARAMCD     label = "Parameter Code"                          length = $8
     AVAL        label = "Analysis Value"                          length = 8
     AVALC       label = "Analysis Value (C)"                      length = $40
     ABLFL       label = "Baseline Record Flag"                    length = $1
     BASE        label = "Baseline Value"                          length = 8
     CHG         label = "Change from Baseline"                    length = 8
     DTYPE       label = "Derivation Type"                         length = $20
     EMDESC      label = "Description of Treatment Emergent"       length = $20
     EGORRES     label = "Result of Finding in Original Units"     length = $100
     EGORRESU    label = "Original Units"                          length = $40
     EGSTRESC    label = "Character Result/ Finding in Std Format" length = $100
     VISIT       label = "Visit Name"                              length = $40
     VISITNUM    label = "Visit Number"                            length = 8
     EGDTC       label = "Date/Time of ECG"                        length = $40

        ;
  set Final;
    keep STUDYID USUBJID SUBJID EGSEQ ASEQ TRTP TRTA ADT ATM ADTM ADY AVISIT
AVISITN APHASE PARAM PARAMCD AVAL AVALC ABLFL BASE CHG DTYPE EMDESC EGORRES
EGORRESU EGSTRESC VISIT VISITNUM EGDTC
        ;
run;
```

Figure 4.3: ADAM.ADEG

AVAL	AVALC	ABLFL	BASE	CHG	DTYPE	EMDESC	EGORRES	EGORRESU	EGSTRESC
52		Y	50			P	50	beats /min	50
56			50	6		T	56	beats /min	56
55			50	5		T	55	beats /min	55
53.66666667			50	3.666666667	AVERAGE	T	55	beats /min	55
	NORMAL	Y				P	NORMAL		NORMAL
	NORMAL					T	NORMAL		NORMAL
	NORMAL					T	NORMAL		NORMAL
150		Y	150			P	150	msec	150
140			150	-10		T	140	msec	140
150			150	0		T	150	msec	150
148.6666667			150	-3.933333333	AVERAGE	T	150	msec	150
83			83			P	83	msec	83
83		Y	83		AVERAGE	P	83	msec	83
100			83	7		T	100	msec	100
93			83	2		T	93	msec	93
409		Y	409			P	409	msec	409
387			409	-22		T	387	msec	387
398.3333333			409	-10.66666667	AVERAGE	T	387	msec	387
395			409	-10		T	395	msec	395
389		Y	389			P	389	msec	389
365			389	-24		T	365	msec	365
376.6666667			389	-12.33333333	AVERAGE	T	365	msec	365
376			389	-13		T	376	msec	376

VISIT	VISITNUM	EGDTC
DAY 1	3	2021-01-02T09:30:00
DAY 2	4	2021-01-03T09:30:00
DAY 3	5	2021-01-04T09:30:00
DAY 3	5	2021-01-04T09:30:00
DAY 1	3	2021-01-02T09:30:00
DAY 2	4	2021-01-03T09:30:00
DAY 3	5	2021-01-04T09:30:00
DAY 1	3	2021-01-02T09:30:00
DAY 2	4	2021-01-03T09:30:00
DAY 3	5	2021-01-04T09:30:00
DAY 3	5	2021-01-04T09:30:00
DAY 1	3	2021-01-02T09:30:00
DAY 1	3	2021-01-02T09:30:00
DAY 2	4	2021-01-03T09:30:00
DAY 3	5	2021-01-04T09:30:00
DAY 1	3	2021-01-02T09:30:00
DAY 2	4	2021-01-03T09:30:00
DAY 2	4	2021-01-03T09:30:00
DAY 3	5	2021-01-04T09:30:00
DAY 1	3	2021-01-02T09:30:00
DAY 2	4	2021-01-03T09:30:00
DAY 2	4	2021-01-03T09:30:00
DAY 3	5	2021-01-04T09:30:00

4.3.2 Lab Test Results Analysis Data Sets (ADLB)

Structure of ADLB

As mentioned in ADEG from the previous section, the structure of ADLB is also BDS and therefore contains one or more records per subject, per analysis parameter, per analysis time point.

Identifier Variables:

- STUDYID: Required variable in character format, the predecessor is *EG.STUDYID*.
- USUBJID: Required variable in character format, the predecessor is *EG.USUBJID*.
- ASEQ: Permissible variable in character format, Analysis Sequence Number, to ensure uniqueness of subject records within an ADaM data set.

Traceability Variables:

- LBSEQ: Permissible variable in character format, the predecessor is *LB.LBSEQ*.

Record-Level Treatment Variables:

- TRTP: Conditional variable in character format, planned treatment, derived from ADSL.
- TRTA: Conditional variable in character format, actual treatment, derived from ADSL.

Timing Variables:

- ADT: Permissible variable in character format, analysis date.
- ATM: Permissible variable in character format, analysis time.
- ADTM: Permissible variable in character format, analysis datetime.
- ADY: Permissible variable in character format, analysis relative day of *AVAL* and/or *AVALC*.
- ASTDT: Permissible variable in character format, analysis start date associated with *AVAL* and/or *AVALC*.
- AVISIT: Conditional variable in character format, the analysis visit description, required if an analysis is done by normal, assigned, or analysis visit.
- AVISITN: Permissible variable in numeric format, representative of *AVISIT*.
- APHASE: Conditional variable in character format, a record-level timing variable that represents the analysis period within the study associated with the record for analysis purposes.
- VISIT: Required variable in character format, the predecessor is *LB.VISIT*.
- VISITNUM: Required variable in numeric format, the predecessor is *LB.VISITNUM*.

Analysis Parameter Variables:

- PARAM: Required variable in character format, the description of the analysis parameter. This variable must include all descriptive and qualifying information relevant to the analysis purpose of the parameter.
- PARAMCD: Required variable in character format, the short name of analysis parameter in *PARAM*.
- AVAL: Conditional variable in numeric format, analysis value, analysis value described by *PARAM*.
- AVALC: Conditional variable in character format, character analysis value described by *PARAM*. *AVALC* can be a character string mapping to *AVAL*.
- BASE: Conditional variable in character format, baseline value. The subject's baseline analysis value for a parameter and baseline definition (*BASETYPE*) if presented. BASE contains the value of *AVAL* copied from a record within the parameter on which *ABLFL*= "Y".
- CHG: Permissible variable in numeric format, change from baseline analysis value. Equal to *AVAL- BASE*. If used for a given *PARAM*, should be populated for all post-baseline records of that *PARAM* regardless of whether that record is used for analysis.

Analysis Descriptor Variables for BDS Data Sets:

- DTYPE: Required variable in character format, derivation type. *DTYPE* is used to denote and must be populated when the value of *AVAL* or *AVALC* has been imputed or derived differently than the other analysis values within the parameter. *DTYPE* is required to be populated *even if AVAL* and *AVALC* are null on the derived record.

Indicator Variables for BDS Data Sets:

- ABLFL: Conditional variable in character format, character indicator to identify the baseline record for the subject, parameter, and baseline type combination. *ABLFL* is required if *BASE* is present in the data set.
- Note: there are other types of variables in the ADaM BDS data sets, such as "Analysis Parameter Criterion Variables", "Toxicity and Range Variables for BDS Data sets", and "BDS population indicator". For a more comprehensive understanding of these variables, please refer to the *ADaM IG 1.2*.

ADLB Specification

Table 4.3: ADLB Specification

Variable	Label	Type	Length	Controlled Terminology	Notes
STUDYID	Study Identifier	char	20		LB.STUDYID
USUBJID	Unique Subject Identifier	char	40		LB.USUBJID
LBSEQ	Sequence Number	num	8		LB.LBSEQ
ASEQ	Analysis Sequence Number	num	8		Sort by USUBJID, PARAMCD, ADT, ATM, DTYPE, LBSEQ, start with 1 for the first value for the first row for each subject and increment by 1 at each successive record.
TRTP	Planned Treatment	char	40		Equal to ADSL.TRT01P
TRTA	Actual Treatment	char	40		Equal to ADSL.TRT01A

(Continued)

Table 4.3: (*Continued*)

Variable	Label	Type	Length	Controlled Terminology	Notes
ADT	Analysis Date	num	8		Equal to numeric date part of LB.LBDTC
ATM	Analysis Time	num	8		
ADTM	Analysis Date and Time	num	8		Equal to LB.LBDTC
ADY	Analysis Relative Day	num	8		Equal to the difference between ADT and ADSL.TRTSDT, if ADT is on or after ADSL.TRTSDT, then add 1.
AVISIT	Analysis Visit	char	40		Equal to "Baseline" if ABLFL is "Y"; Equal to "Follow-Up" if VISIT is "Follow Up"; else equal to VISIT.
AVISITN	Analysis Visit (N)	num	8		Equal to 0, if AVISIT is "Baseline"; equal to 100, if AVISIT is "Follow-Up"; else equal to X (depending on which visit it is and what the sponsor convention is).
APHASE	Phase	char	40		Equal to "Screening" when the assessment is on or before the first dose of study drug; equal to "Treatment" when the assessment is after the first dose of study drug. Equal to "Follow-Up" when the assessment is after the last dose date.

(*Continued*)

Table 4.3: (*Continued*)

Variable	Label	Type	Length	Controlled Terminology	Notes
PARAM	Parameter	char	40		Concatenate LB.LBTEST with LB.LBORRESU
PARAMCD	Parameter Code	char	8		LB.LBTESTCD
AVAL	Analysis Value	num	8		Equal to LBSTRESN if LBSTRESN is not null; if LBSTRESN is null and LBSTRESC is expressed with "<" or "<=" or ">" or ">=", then set to the numeric part of LBSTRESC.
AVALC	Analysis Value (C)	char	40		Equal to LB.LBSTRESC.
ABLFL	Baseline Record Flag	char	1		Only populate when DTYPE is "AVERAGE", equal to "Y" for the most recent nonmissing measurement before the first dose of study drug.
BASE	Baseline Value	num	8		Only populate baseline and post-baseline (EMDESC in ("P","T"))when ABLFL is "Y", equal to AVAL, not populate when EMDESC is "A".
CHG	Change from Baseline	num	8		Only populate post-baseline (EMDESC in "T") when AVAL and BASE are not missing , equal to (AVAL - BASE).
DTYPE	Derivation Type	char	20		Equal to "IMPUTE" when AVAL is derived value.

(*Continued*)

Table 4.3: (*Continued*)

Variable	Label	Type	Length	Controlled Terminology	Notes
EMDESC	Description of Treatment Emergent	char	20		Equal to "P" when the assessment is on or before the first dose of study drug (ADT<=ADSL. TRTSDT and ATM is missing. If ADT and ATM is not missing, then ADTM<=ADSL. TRT01SDTM); equal to "T" when the assessment is after the first dose of study drug (ADT > ADSL. TRTSDT and ATM is missing, or if ADT and ATM is not missing then ADTM>ADSL. TRT01SDTM); otherwise, equal to "A" when the assessment is after the end of the treatment
LBORRES	Result of Finding in Original Units	char	100		LB.LBORRES
LBORRESU	Original Units	char	40		LB.ORRESU
LBORNRLO	Reference Range Lower Limit in Orig Unit	char	40		LB.LBORNRLO
LBORNRHI	Reference Range Higher Limit in Orig Unit	char	40		LB.LBORNRHI

(*Continued*)

Table 4.3: (*Continued*)

Variable	Label	Type	Length	Controlled Terminology	Notes
LBSTNRLO	Reference Range Lower Limit-Std Units	char	40		LB.LBSTNRLO
LBSTNRHI	Reference Range Upper Limit-Std Units	num	8		LB.LBSTNRHI
LBSTRESC	Character Result/ Finding in Std Format	char	40		LB.LBSTRESC
VISIT	Visit Name	char	40		LB.VISIT
VISITNUM	Visit Number	num	8		LB.VISITNUM
LBDTC	Date/Time of ECG	char	40		LB.LBDTC; equal to null for the average records

ADLB Programming

In order to create the ADLB domain, we start by merging SDTM.LB and ADAM.ADSL. Please see the code below.

Note: For every domain, the corresponding SAS name is domain.sas. For example, ADLB is created by adlb.sas, so all code in this section belongs to adlb.sas.

```
/*Begin writing SAS program merge SDTM.LB and ADAM.ADSL.sas*/

data ADLB1;
  merge SDTM.LB(in=a) ADAM.ADSL(in=b drop=STUDYID);
  by USUBJID;
  if a and b;
run;

data ADLB2;
  set ADLB1;
  /*Derive ADT, ATM, ADTM*/

  length PARAMCD $8. PARAM AVALC $40. AVAL 8;
  if length(LBDTC)=10 then do;
    ADT=input(LBDTC,yymmdd10.);
    ATM=.;
```

```
   ADTM=.;
 end;
 if length(LBDTC)>10 then do;
   ADTM=input(LBDTC,is8601dt.);
   ADT=datepart(ADTM);
   ATM=timepart(ADTM);
 end;
 format ADTM is8601dt. ADT yymmdd10. ATM time5.;
 /*Derive ADY*/
 if nmiss(ADT,TRTSDT)=0 then ADY=ADT-TRTSDT+(ADT>=TRTSDT);


 /*Derive APHASE, EMDESC*/


 if (ADT<=TRTSDT and ATM=.) or (ADT^=. and ATM^=. and ADTM<=TRT01SDTM) then
do;
    APHASE="Screening";
    EMDESC="P";
 end;
 if (ADT > TRTSDT and ATM =.) or (ADT^=. and ATM^=. and ADTM>TRT01SDTM) then
do;
    APHASE="Treatment";
    EMDESC="T";
 end;
  if (ADT > TRTEDT and ATM =.) or (ADT^=. and ATM^=. and ADTM>TRT01EDTM)
then do ;
    APHASE="Follow-Up";
    EMDESC="A";
 end;
 /*Derive PARAM, PARAMCD*/
 PARAM=strip(LBTEST)||" ("||strip(LBORRESU)||")";
 PARAMCD=strip(LBTESTCD);

 /*Derive AVAL, AVALC, DTYPE*/

 if ^missing(LBSTRESN) then do;
   AVAL=LBSTRESN;
   DTYPE="";
 end;

 else if LBSTRESN=. and ^missing(LBSTRESC) then do;
  if index(LBSTRESC,"<") or index(LBSTRESC,"<=") then do;
     AVAL=input(compress(LBSTRESC,"<="),best.);
     DTYPE="IMPUTE";
  end;
 if index(LBSTRESC,">") or index(LBSTRESC,">=") then do;
    AVAL=input(compress(LBSTRESC,">="),best.);
    DTYPE="IMPUTE";
  end;
 end;
 AVALC=strip(LBSTRESC);
```

```
/*Derive TRTP, TRTA*/
TRTP=TRT01P;
TRTA=TRT01A;

run;
```

The following code is to count *NUMBER* for the purpose of deriving *ABLFL* and, later, merging *ABLFL* variable back to ADLB2 based on the same *NUMBER*.

```
/*Derive ABLFL*/
data ADLB2;
  set ADLB2;
  NUMBER=_n_;
run;
proc sort data=ADLB2;
  by USUBJID PARAMCD ADT ADTM;
run;
/*Filter the condition of baseline flag */


data BASE;
  set ADLB2(where=(EMDESC="P" and (AVAL ne . or AVALC ne '')  and
(.<ADT<=TRTSDT)));
  by USUBJID PARAMCD ADT ADTM;
run;

/*if the last PARAMCD then ABLFL sets to "Y" */

data ABLFL;
   set BASE;
   by USUBJID PARAMCD ADT ADTM;
  if last.PARAMCD; ABLFL="Y";    ❶
run;
/*Left Join ADLB2 with ABLFL */

proc sql;
      create table ADLB3 as select a.*,b.ABLFL from ADLB2 as a left join
ABLFL as b on a.NUMBER=b.NUMBER;      ❷
quit;
```

❶ Refer to ADAM specification. *ABLFL* is equal to "Y" for the most recent nonmissing measurement before the first dose of study drug.

❷ Merge *ABLFL* from ADEG10 back to ADEG11 based on the same *NUMBER*.

```
/*Derive variable Base */
proc sql;
   create table ADLB4 as select a.*,b.AVAL as BASE from ADLB3 as a
    left join ADLB3(where=(ABLFL='Y')) as b on a.USUBJID=b.USUBJID and
a.PARAMCD=b.PARAMCD;
quit;
/*Derive variable CHG*/
```

```
data ADLB5;
  set ADLB4;
  if n(AVAL,BASE)=2 and ABLFL ne "Y"  then CHG=AVAL-BASE;
  if ABLFL^="Y" and EMDESC="P" then do;
    CHG=.;
  end;  ❸
run;
```

❸ Derive *CHG* (change from baseline), which is only populated for the post-baseline records.

```
/*Derive variable AVISIT and AVISITN*/

data ADLB6;
  set ADLB5;
  length AVISIT $40. AVISITN 8. ;if ABLFL="Y" then do;
    AVISIT="Baseline";
    AVISITN=0;
end;
else if index(VISIT,"FOLLOW-UP") then do;
    AVISIT="Follow-up";
    AVISITN=100;
end;
else do;
    AVISIT=strip(VISIT);
    AVISITN=input(compress(AVISIT,,"kd"),best.);  ❹
end;

run;
```

❹   Derive *AVISIT, AVISITN*. We use the SAS function compress () with "kd" to only keep the numeric value for *AVISIT*.

```
/*Derive ASEQ*/
proc sort data=ADLB6;
   by USUBJID PARAMCD ADT ATM DTYPE LBSEQ;
run;
data Final;
  set ADLB6;
  by USUBJID PARAMCD ADT ATM DTYPE LBSEQ;
  if first.USUBJID then ASEQ = 0;           ❺
   ASEQ+1;
   output;
run;
```

❺   Create variable *ASEQ* variable. This variable is used to uniquely record analysis sequence number of subjects within a domain.

```
libname ADAM ".../directory";
data ADAM.ADLB(label="Lab Test Results Analysis Data sets");
```

```
/*Assign variable attributes such as label and length to conform with ADAM.
ADSL Specification (these will also be the same attributes as the ADAM IG).*/

attrib
    STUDYID     label = "Study Identifier"                      length = $20
    USUBJID     label = "Unique Subject Identifier"             length = $40
    SUBJID      label = "Subject Identifier for the Study"      length = $20
    LBSEQ       label = "Sequence Number"                       length = 8
    ASEQ        label = "Analysis Sequence Number"              length = 8
    TRTP        label = "Planned Treatment"                     length = $40
    TRTA        label = "Actual Treatment"                      length = $40
    ADT         label = "Analysis Date"                         length = 8
    ATM         label = "Analysis Time"                         length = 8
    ADTM        label = "Analysis Date and Time"                length = 8
    ADY         label = "Analysis Relative Day"                 length = 8
    AVISIT      label = "Analysis Visit"                        length = $40
    AVISITN     label = "Analysis Visit (N)"                    length = 8
    APHASE      label = "PHASE"                                 length = $40
    PARAM       label = "Parameter"                             length = $40
    PARAMCD     label = "Parameter Code"                        length = $8
    AVAL        label = "Analysis Value"                        length = 8
    AVALC       label = "Analysis Value (C)"                    length = $40
    ABLFL       label = "Baseline Record Flag"                  length = $1
    BASE        label = "Baseline Value"                        length = 8
    CHG         label = "Change from Baseline"                  length = 8
    DTYPE       label = "Derivation Type"                       length = $20
    EMDESC      label = "Description of Treatment Emergent"     length = $20
    LBORRES     label = "Result of Finding in Original Units"   length = $100
    LBORRESU    label = "Original Units"                        length = $40
    LBORNRLO    label = "Reference Range Lower Limit in Orig Unit" length = $40
    LBORNRHI    label = "Reference Range Higher Limit in Orig Unit"length = $40
    LBSTNRLO    label = "Reference Range Lower Limit-Std Units"  length = 8
    LBSTNRHI    label = "Reference Range Upper Limit-Std Units"  length = 8
    LBSTRESC    label = "Character Result/ Finding in Std Format" length = $100
    VISIT       label = "Visit Name"                            length = $40
    VISITNUM    label = "Visit Number"                          length = 8
    LBDTC       label = "Date/Time of Lab"                      length = $40
        ;
  set Final;
    keep STUDYID USUBJID SUBJID LBSEQ ASEQ TRTP TRTA ADT ATM ADTM ADY AVISIT
AVISITN APHASE PARAM PARAMCD AVAL AVALC ABLFL BASE CHG DTYPE EMDESC LBORRES
LBORRESU LBORNRLO LBSTNRHI LBSTNRLO LBSTNRHI LBSTRESC VISIT VISITNUM
LBDTC
        ;
run;
```

Figure 4.4: ADAM.ADLB

	STUDYID	USUBJID	SUBJID	LBSEQ	ASEQ	TRTP	TRTA
1	ABC-001	ABC-001-001-001	001	1	1	TRTA	TRTA
2	ABC-001	ABC-001-001-001	001	2	2	TRTA	TRTA
3	ABC-001	ABC-001-001-002	002	1	1	TRTA	TRTA
4	ABC-001	ABC-001-001-002	002	2	2	TRTA	TRTA
5	ABC-001	ABC-001-001-003	003	1	1	TRTA	TRTA
6	ABC-001	ABC-001-001-003	003	2	2	TRTA	TRTA
7	ABC-001	ABC-001-001-004	004	1	1	TRTA	TRTA
8	ABC-001	ABC-001-001-004	004	2	2	TRTA	TRTA
9	ABC-001	ABC-001-001-005	005	1	1	TRTA	TRTA
10	ABC-001	ABC-001-001-005	005	2	2	TRTA	TRTA
11	ABC-001	ABC-001-001-006	006	1	1	TRTA	TRTA
12	ABC-001	ABC-001-001-006	006	2	2	TRTA	TRTA
13	ABC-001	ABC-001-001-007	007	1	1	TRTA	TRTA
14	ABC-001	ABC-001-001-007	007	2	2	TRTA	TRTA
15	ABC-001	ABC-001-001-008	008	1	1	TRTA	TRTA
16	ABC-001	ABC-001-001-008	008	2	2	TRTA	TRTA
17	ABC-001	ABC-001-001-009	009	1	1	TRTA	TRTA
18	ABC-001	ABC-001-001-009	009	2	2	TRTA	TRTA
19	ABC-001	ABC-001-001-010	010	1	1	TRTA	TRTA
20	ABC-001	ABC-001-001-010	010	2	2	TRTA	TRTA

ADT	ATM	ADTM	ADY	AVISIT	AVISITN	APHASE	PARAM	PARAMCD
2021-01-02	9:45	2021-01-02T09:45:00	1	Baseline	0	Screening	ALKALINE PHOSPHATASE (U/L)	ALP
2021-01-02	9:45	2021-01-02T09:45:00	1	Baseline	0	Screening	CALCIUM (MG/DL)	CA
2021-01-02	9:45	2021-01-02T09:45:00	1	Baseline	0	Screening	ALKALINE PHOSPHATASE (U/L)	ALP
2021-01-02	9:45	2021-01-02T09:45:00	1	Baseline	0	Screening	CALCIUM (MG/DL)	CA
2021-01-02	9:45	2021-01-02T09:45:00	1	Baseline	0	Screening	ALKALINE PHOSPHATASE (U/L)	ALP
2021-01-02	9:45	2021-01-02T09:45:00	1	Baseline	0	Screening	CALCIUM (MG/DL)	CA
2021-01-02	9:45	2021-01-02T09:45:00	1	Baseline	0	Screening	ALKALINE PHOSPHATASE (U/L)	ALP
2021-01-02	9:45	2021-01-02T09:45:00	1	Baseline	0	Screening	CALCIUM (MG/DL)	CA
2021-01-02	9:45	2021-01-02T09:45:00	1	Baseline	0	Screening	ALKALINE PHOSPHATASE (U/L)	ALP
2021-01-02	9:45	2021-01-02T09:45:00	1	Baseline	0	Screening	CALCIUM (MG/DL)	CA
2021-01-02	9:45	2021-01-02T09:45:00	1	Baseline	0	Screening	ALKALINE PHOSPHATASE (U/L)	ALP
2021-01-02	9:45	2021-01-02T09:45:00	1	Baseline	0	Screening	CALCIUM (MG/DL)	CA
2021-01-02	9:45	2021-01-02T09:45:00	1	Baseline	0	Screening	ALKALINE PHOSPHATASE (U/L)	ALP
2021-01-02	9:45	2021-01-02T09:45:00	1	Baseline	0	Screening	CALCIUM (MG/DL)	CA
2021-01-02	9:45	2021-01-02T09:45:00	1	Baseline	0	Screening	ALKALINE PHOSPHATASE (U/L)	ALP
2021-01-02	9:45	2021-01-02T09:45:00	1	Baseline	0	Screening	CALCIUM (MG/DL)	CA
2021-01-02	9:45	2021-01-02T09:45:00	1	Baseline	0	Screening	ALKALINE PHOSPHATASE (U/L)	ALP
2021-01-02	9:45	2021-01-02T09:45:00	1	Baseline	0	Screening	CALCIUM (MG/DL)	CA
2021-01-02	9:45	2021-01-02T09:45:00	1	Baseline	0	Screening	ALKALINE PHOSPHATASE (U/L)	ALP
2021-01-02	9:45	2021-01-02T09:45:00	1	Baseline	0	Screening	CALCIUM (MG/DL)	CA

AVAL	AVALC	ABLFL	BASE	CHG	DTYPE	EMDESC	LBORRES
62	62	Y	62			P	62
2.25	2.25	Y	2.25			P	9
65	65	Y	65			P	65
2.375	2.375	Y	2.375			P	9.5
66	66	Y	66			P	66
2.3	2.3	Y	2.3			P	9.2
67	67	Y	67			P	67
2.35	2.35	Y	2.35			P	9.4
64	64	Y	64			P	64
2.375	2.375	Y	2.375			P	9.5
63	63	Y	63			P	63
2.3	2.3	Y	2.3			P	9.2
66	66	Y	66			P	66
2.325	2.325	Y	2.325			P	9.3
60	60	Y	60			P	60
2.25	2.25	Y	2.25			P	9
62	62	Y	62			P	62
2.3	2.3	Y	2.3			P	9.2
63	63	Y	63			P	63
2.3	2.3	Y	2.3			P	9.2

LBORRESU	LBORNRLO	LBSTNRLO	LBSTNRHI	LBSTRESC	VISIT	VISITNUM	LBDTC
U/L	50	100	100	62	DAY 1	3	2021-01-02T09:45
MG/DL	8.5	2.5	2.5	2.25	DAY 1	3	2021-01-02T09:45
U/L	50	100	100	65	DAY 1	3	2021-01-02T09:45
MG/DL	8.5	2.5	2.5	2.375	DAY 1	3	2021-01-02T09:45
U/L	50	100	100	56	DAY 1	3	2021-01-02T09:45
MG/DL	8.5	2.5	2.5	2.3	DAY 1	3	2021-01-02T09:45
U/L	50	100	100	67	DAY 1	3	2021-01-02T09:45
MG/DL	8.5	2.5	2.5	2.35	DAY 1	3	2021-01-02T09:45
U/L	50	100	100	64	DAY 1	3	2021-01-02T09:45
MG/DL	8.5	2.5	2.5	2.375	DAY 1	3	2021-01-02T09:45
U/L	50	100	100	63	DAY 1	3	2021-01-02T09:45
MG/DL	8.5	2.5	2.5	2.3	DAY 1	3	2021-01-02T09:45
U/L	50	100	100	66	DAY 1	3	2021-01-02T09:45
MG/DL	8.5	2.5	2.5	2.125	DAY 1	3	2021-01-02T09:45
U/L	50	100	100	50	DAY 1	3	2021-01-02T09:45
MG/DL	8.5	2.5	2.5	2.25	DAY 1	3	2021-01-02T09:45
U/L	50	100	100	62	DAY 1	3	2021-01-02T09:45
MG/DL	8.5	2.5	2.5	2.3	DAY 1	3	2021-01-02T09:45
U/L	50	100	100	63	DAY 1	3	2021-01-02T09:45
MG/DL	8.5	2.5	2.5	2.3	DAY 1	3	2021-01-02T09:45

4.4 Occurrence Data Structure (OCCDS)

The occurrence data structure describes the general data configuration and content found in occurrence analysis. Occurrence analysis is the counting of subjects with a given record or term. Both Events and Interventions domains are OCCDS data structures. For example, Adverse Event and Concomitant Medications are OCCDS and are described in this chapter. The structure of occurrence analysis data sets is one record per subject per event (or intervention such as in ADCM) per time point and is based on the corresponding SDTM domain.

4.4.1 Adverse Event Analysis Data Set (ADAE)

Structure of ADAE

The structure of ADAE is OCCDS and therefore one record per subject per event per time point and is based on the corresponding SDTM AE domain.

Identifier Variables:

- STUDYID: Required variable in character format, the predecessor is *AE.STUDYID*.
- USUBJID: Required variable in character format, the predecessor is *AE.USUBJID*.

Traceability Variables:

- AESEQ: Permissible variable in character format, the predecessor is *AE.EGSEQ*.

Record-Level Treatment Variables:

- TRTP: Conditional variable in character format, planned treatment, derived from ADSL.
- TRTA: Conditional actual treatment, derived from ADSL.

Dictionary Coding and Categorization Variables:

- AETERM: Required variable in character format, the predecessor is *AE.AETERM*.
- AEDECOD: Conditional variable in character format, the predecessor is *AE.AEDECOD*.
- AELLT: Conditional variable in character format, the predecessor is *AE.AELLT*.
- AELLTCD: Permissible variable in numeric format, the predecessor is *AE.AELLTCD*.
- AEHLT: Conditional variable in character format, the predecessor is *AE.AEHLT*.
- AEHLTCD: Permissible variable in numeric format, the predecessor is *AE.AEHLTCD*.
- AEHLGT: Conditional variable in character format, the predecessor is *AE.AEHLGT*.
- AEPTCD: Permissible variable in character format, the predecessor is *AE.AEPTCD*.

Timing Variables:

- AESTDTC: Permissible variable in character format, the predecessor is *AE.AESTDTC*.
- ASTDT: Conditional variable in numeric format, conditional on whether start date is pertinent for study and is populated in SDTM.
- AEENDTC: Conditional variable in character format, the predecessor is *AE.AEENDTC*.
- AENDT: Conditional variable in numeric format, conditional on whether end date is pertinent for study and is populated in SDTM.
- ASTDY: Conditional variable in numeric format, conditional on whether analysis date is pertinent for study and is populated in SDTM.

Indicator Variables for OCCDS Data Sets:

- TRTEMFL: Conditional variable in character format, character indicator of whether the observation occurred while the subject was on treatment.
- PREFL: Conditional variable in character format, character indicator of whether the observation occurred before the subject started treatment.
- FUPFL: Conditional variable in character format, character indicator of whether the observation occurred after the subject was on treatment.

Descriptive Variables:

- AESER: Required variable in character format, the predecessor is *AE.AESER*.
- AEACN: Permissible variable in character format, the predecessor is *AE.AEACN*.
- AEREL: Permissible variable in character format, the predecessor is *AE.AEREL*.
- AEOUT: Permissible variable in character format, the predecessor is *AE.AEOUT*.
- AESCONG: Permissible variable in character format, the predecessor is *AE.AESCONG*.
- AESDISAB: Permissible variable in character format, the predecessor is *AE.AESDISAB*.
- AESDTH: Permissible variable in character format, the predecessor is *AE.AESDTH*.
- AESHOSP: Permissible variable in character format, the predecessor is *AE.AESHOSP*.
- AECONTRT: Permissible variable in character format, the predecessor is *AE.AECONTRT*.
- ARTOXGR: Permissible variable in character format, the predecessor is *AE.ARTOXGR*.

Note: there are other types of variables in the ADaM OCCDS data sets, such as "Occurrence flags", "Treatment/Dose Variables", "Standardized MedDRA Query Variables", and "Original or

Prior Coding Variables". Please refer to *ADaM IG 1.2* for a more comprehensive understanding of these variables.

ADAE Specification

Table 4.4: ADAE Specification

Variable	Label	Type	Length	Controlled Terminology	Notes
STUDYID	Study Identifier	char	20		AE.STUDYID
USUBJID	Unique Subject Identifier	char	40		AE.USUBJID
ASEQ	Sequence Number	num	8		AE.AESEQ
TRTP	Planned Treatment	char	40		Equal to ADSL.TRT01P
TRTA	Actual Treatment	char	40		Equal to ADSL.TRT01A
AETERM	Reported Term for the Adverse Event	char	200		AE.AETERM
AEDECOD	Dictionary-Derived Term	char	200		AE.AEDECOD
AEBODSYS	Body System or Organ Class	char	20		AE.AEBODSYS
AELLT	Lowest Level Term	char	100		AE.AELLT
AELLTCD	Lowest Level Term Code	num	8		AE.AELLTCD
AEHLT	High-Level Term	char	200		AE.AEHLT
AEHLTCD	High-Level Term Code	num	8		AE.AEHLTCD
AEHLGT	High-Level Group Term	char	200		AE.AEHLGT
AEHLGTCD	High-Level Group Term Code	num	8		AE.AEHLGTCD

(Continued)

Table 4.4: (*Continued*)

Variable	Label	Type	Length	Controlled Terminology	Notes
AEPTCD	Preferred Term Code	num	8		AE.AEPTCD
AESTDTC	Start Date/ Time of Adverse Event	char	20		AE.AESTDTC
ASTDT	Analysis Start Date	num	8		Numeric part of AESTDTC
ASTDTF	Analysis Start Date Imputation Flag	char	1		Equal to "D" (Day) if only the day is imputed; equal to "M" (Month) if only the month is imputed.
AEENDTC	End Date/Time of Adverse Event	char	20		AE.AEENDTC
AENDT	Analysis Start Date	num	8		Numeric part of AEENDTC
ASTDY	Analysis Start Relative Day	num	8		Equal to the difference between ASTDT and ADSL. TRTSDT, if ASTDT is on or after ADSL.TRTSDT, then add 1.
AESTDY	Study Day of Start of Adverse Event	num	8		AE.AESTDY
AEENDY	Study Day of End of Adverse Event	num	8		AE.AEENDY
TRTEMFL	Treatment Emergent Analysis Flag	char	1		Equal to "Y" if the AE starts on or after the first dose date of study drug through the end of the treatment emergent period (ADSL.TRTSDT <= ASTDT <= ADSL. TSFENDT)

(*Continued*)

Table 4.4: (*Continued*)

Variable	Label	Type	Length	Controlled Terminology	Notes
PREFL	Pre-treatment Flag	char	1		Equal to "Y" if the AE starts before the first dose of study drug; otherwise set to null. (ASTDT<ADSL.TRTSDT)
FUPFL	Follow-Up Flag	char	1		Equal to "Y" if the AE starts on or after the end of the treatment emergent period (ASTDT >= ADSL.TSFENDT)
EMDESC	Description of Treatment Emergent	char	20		Equal to "T" when TRTEMFL is "Y"; equal to "P" when PREFL is "Y", otherwise is equal to "A".
AESER	Serious Event	char			AE.AESER
AEACN	Action Taken with Study Treatment	char			AE.AEACN
AEREL	Causality	char			AE.AEREL
AEOUT	Outcome of Adverse Event	char			AE.AEOUT
AESCONG	Congenital Anomaly or Birth Defect	char			AE.AESCONG
AESDISAB	Persist or Significant Disability	char			AE.AESDISAB
AESDTH	Results in Death	char			AE.AESDTH
AESHOSP	Requires or Prolongs Hospitalization	char			AE.AESHOSP
AECONTRT	Concomitant or Additional Treatment Given	char			AE.AECONTRT
AETOXGR	Standard Toxicity Grade	char			AE.AETOXGR

ADAE Programming

In order to create the ADAE domain, we start by merging SDTM.AE and ADAM.ADSL. Please see code below.

Note: For every domain, the corresponding SAS name is domain.sas. For example, ADAE is created by adae.sas, so all code in this section belongs to adae.sas.

```
/*Begin writing SAS program - merge SDTM.AE and ADAM.ADSL.sas'/
data ADAE1;
  merge SDTM.AE(in=a) ADAM.ADSL(in=b drop=STUDYID);       ❶
by USUBJID;
if a and b;
run;
```

❶ Merge SDTM.AE and ADAM.ADSL to get core variables from ADAM.ADSL.

```
/*Derive ASTDT, AENDT */

data FINAL;
  set ADAE1;
  length TRTEMFL PREFL ASTDTF $1 ASTDT AENDT 8. ;
  /*Derive TRTEMFL, PREFL, FUPFL, EMDESC*/

  ASTDT=input(substr(AESTDTC,1,10),yymmdd10.);
  AENDT=input(substr(AEENDTC,1,10),yymmdd10.);
  format ASTDT AENDT yymmdd10.;

  /*Derive TRTEMFL, PREFL, FUPFL, EMDESC*/
  if TRTSDT <= ASTDT <= TRTEDT then do;
    TRTEMFL = "Y";
    EMDESC = "T";
  end;

  if . < ASTDT < TRTSDT then do;
    PREFL = "Y";
    EMDESC = "P";
    end;
  else if ASTDT > TRTEDT and TRTEDT^=. then do;
    FUPFL = "Y";
    EMDESC= "A";
  end;
  if nmiss(ASTDT,TRTSDT)=0 then ASTDY=ASTDT-TRTSDT+(ASTDT>=TRTSDT);

  /*Derive ASTDTF*/
```

```
   if length(AESTDTC) >= 10 then ASTDTF="";
   else if length(AESTDTC)=7 then ASTDTF="D";
   else if length(AESTDTC)=4 then ASTDTF="M";

   /*Derive TRTP, TRTA*/
   TRTA=TRT01A;
   TRTP=TRT01P;

run;

libname ADAM ".../directory";

data ADAM.ADAE(label="Adverse Events Analysis Data Set");

/*Assign variable attributes such as label and length to conform with ADAM.
ADSL Specification (these will also be the same attributes as the ADAM IG).*/

attrib
   STUDYID     label = "Study Identifier"                          length = $20
   USUBJID     label = "Unique Subject Identifier"                 length = $40
   AESEQ       label = "Sequence Number"                           length = 8
   AETERM      label = "Reported Term for the Adverse Event"       length = $200
   AELLT       label = "Lowest Level Term"                         length = $100
   AELLTCD     label = "Lowest Level Term Code"                    length = 8
   AEDECOD     label = "Dictionary-Derived Term"                   length = $200
   AEPTCD      label = "Preferred Term Code"                       length = 8
   ASTDT       label = "Analysis Start Date"                       length = 8
   ASTDTF      label = "Analysis Start Date Imputation Flag"       length = $1
   AEENDTC     label = "End Date/Time of Adverse Event"            length = $20
   AENDT       label = "Analysis Start Date"                       length = 8
   ASTDY       label = "Analysis Start Relative Day"               length = 8
   TRTEMFL     label = "Treatment Emergent Analysis Flag"          length = $1
   PREFL       label = "Pre-treatment Flag"                        length = $1
   FUPFL       label = "Follow-Up Flag"                            length = $1
   EMDESC      label = "Description of Treatment Emergent "        length = $20
   AEHLT       label = "High-Level Term"                           length = $200
   AEHLTCD     label = "High-Level Term Code"                      length = 8
   AEHLGT      label = "High-Level Group Term"                     length = $200
   AEHLGTCD    label = "High-Level Group Term Code"                length = 8
   AEBODSYS    label = "Body System or Organ Class"                length = $20
   AESER       label = "Serious Event"                             length = $2
   AEACN       label = "Action Taken with Study Treatment"         length = $50
   AEREL       label = "Causality"                                 length = $50
   AEOUT       label = "Outcome of Adverse Event"                  length = $50
   AESCONG     label = "Congenital Anomaly or Birth Defect"        length = $2
   AESDISAB    label = "Persist or Significant Disability"         length = $2
   AESDTH      label = "Results in Death"                          length = $2
```

```
        ;
    set FINAL;

    keep STUDYID USUBJID AESEQ AETERM AELLT AELLTCD AEDECOD AEPTCD ASTDT
ASTDTF AEENDTC AENDT ASTDY TRTEMFL PREFL FUPFL EMDESC AEHLT AEHLTCD AEHLGT
AEHLGTCD AEBODSYS AESER AEACN AEREL AEOUT AESCONG AESDISAB AESDTH
        ;
run;
```

Figure 4.5: ADAM.ADAE

	STUDYID	USUBJID	AESEQ	AETERM	AELLT	AELLTCD	AEDECOD	AEPTCD
1	ABC-001	ABC-001-001-001	1	HEADACHE	HEADACHE	10019211	HEADACHES	10019211
2	ABC-001	ABC-001-001-002	1	SORE NECK	NECK PAIN	10028836	NECK PAIN	10028836
3	ABC-001	ABC-001-001-003	1	SORE NECK	NECK PAIN	10028836	NECK PAIN	10028836
4	ABC-001	ABC-001-001-004	1	SORE NECK	NECK PAIN	10028836	NECK PAIN	10028836
5	ABC-001	ABC-001-001-005	1	SORE NECK	NECK PAIN	10028836	NECK PAIN	10028836
6	ABC-001	ABC-001-001-006	1	SORE NECK	NECK PAIN	10028836	NECK PAIN	10028836
7	ABC-001	ABC-001-001-007	1	SORE NECK	NECK PAIN	10028836	NECK PAIN	10028836
8	ABC-001	ABC-001-001-008	1	SORE NECK	NECK PAIN	10028836	NECK PAIN	10028836
9	ABC-001	ABC-001-001-009	1	SORE NECK	NECK PAIN	10028836	NECK PAIN	10028836
10	ABC-001	ABC-001-001-010	1	HEADACHE	HEADACHE	10019211	HEADACHES	10019211

ASTDT	ASTDTF	AEENDTC	AENDT	ASTDY	TRTEMFL	PREFL	FUPFL	EMDESC	AEHLT
2021-01-05	2021-01-05	2021-01-05	4		Y	A			HEADACHES NEC
2021-01-07	2021-01-07	2021-01-07	6		Y	A			MUSCULOSKELETAL AND CONNECTIVE TISSUE DISORDERS
2021-01-07	2021-01-07	2021-01-07	6		Y	A			MUSCULOSKELETAL AND CONNECTIVE TISSUE DISORDERS
2021-01-07	2021-01-07	2021-01-07	6		Y	A			MUSCULOSKELETAL AND CONNECTIVE TISSUE DISORDERS
2021-01-07	2021-01-07	2021-01-07	6		Y	A			MUSCULOSKELETAL AND CONNECTIVE TISSUE DISORDERS
2021-01-07	2021-01-07	2021-01-07	6		Y	A			MUSCULOSKELETAL AND CONNECTIVE TISSUE DISORDERS
2021-01-07	2021-01-07	2021-01-07	6		Y	A			MUSCULOSKELETAL AND CONNECTIVE TISSUE DISORDERS
2021-01-07	2021-01-07	2021-01-07	6		Y	A			MUSCULOSKELETAL AND CONNECTIVE TISSUE DISORDERS
2021-01-09	2021-01-09	2021-01-09	8		Y	A			MUSCULOSKELETAL AND CONNECTIVE TISSUE DISORDERS
2021-01-09	2021-01-09	2021-01-09	8		Y	A			HEADACHES NEC

AEHLTCD	AEHLGT	AEHLGTCD	AEBODSYS	AESER	AEACN	AEREL
10019233	HEADACHES	10019231	NERVOUS SYSTEM DISOR	N	DRUG NOT CHANGED	N
10068757	MUSCULOSKELETAL AND CONNECTIVE TISSUE DISORDERS NEC	10028393	MUSCULOSKELETAL AND	N	DRUG NOT CHANGED	
10068757	MUSCULOSKELETAL AND CONNECTIVE TISSUE DISORDERS NEC	10028393	MUSCULOSKELETAL AND	N	DRUG NOT CHANGED	N
10068757	MUSCULOSKELETAL AND CONNECTIVE TISSUE DISORDERS NEC	10028393	MUSCULOSKELETAL AND	N	DRUG NOT CHANGED	
10068757	MUSCULOSKELETAL AND CONNECTIVE TISSUE DISORDERS NEC	10028393	MUSCULOSKELETAL AND	N	DRUG NOT CHANGED	N
10068757	MUSCULOSKELETAL AND CONNECTIVE TISSUE DISORDERS NEC	10028393	MUSCULOSKELETAL AND	N	DRUG NOT CHANGED	N
10068757	MUSCULOSKELETAL AND CONNECTIVE TISSUE DISORDERS NEC	10028393	MUSCULOSKELETAL AND	N	DRUG NOT CHANGED	
10068757	MUSCULOSKELETAL AND CONNECTIVE TISSUE DISORDERS NEC	10028393	MUSCULOSKELETAL AND	N	DRUG NOT CHANGED	N
10068757	MUSCULOSKELETAL AND CONNECTIVE TISSUE DISORDERS NEC	10028393	MUSCULOSKELETAL AND	N	DRUG NOT CHANGED	
10019233	HEADACHES	10019231	NERVOUS SYSTEM DISOR	N	DRUG NOT CHANGED	Y

AEOUT	AESCONG	AESDISAB	AESDTH
RECOVERD OR RESOLVED	N	N	N
RECOVERD OR RESOLVED	N	N	N
RECOVERD OR RESOLVED	N	N	N
RECOVERD OR RESOLVED	N	N	N
RECOVERD OR RESOLVED	N	N	N
RECOVERD OR RESOLVED	N	N	N
RECOVERD OR RESOLVED	N	N	N
RECOVERD OR RESOLVED	N	N	N
RECOVERD OR RESOLVED	N	N	N
RECOVERD OR RESOLVED	N	N	N

4.4.2 Concomitant Medications Analysis Data Set (ADCM)

Structure of ADCM

As mentioned earlier, the structure of ADCM is OCCDS and therefore one record per subject per event per time point and is based on the corresponding SDTM CM domain.

Identifier Variables:

- STUDYID: Required variable in character format, the predecessor is CM.STUDYID.
- USUBJID: Required variable in character format, the predecessor is CM.USUBJID.

Traceability Variables:

- CMSEQ: Permissible variable in character format, the predecessor is CM.EGSEQ.

Record-Level Treatment Variables:

- TRTP: Conditional variable in character format, planned treatment, derived from ADSL.
- TRTA: Conditional variable in character format, actual treatment, derived from ADSL.

Dictionary Coding and Categorization Variables:

- CMTERM: Required variable in character format, the predecessor is *CM.CMTERM*.
- CMDECOD: Conditional variable in character format, the predecessor is *CM.CMDECOD*.

Timing Variables:

- CMSTDTC: Permissible variable in character format, the predecessor is *CM.CMSTDTC*.
- ASTDT: Conditional variable in numeric format, conditional on whether start date is pertinent for study and is populated in SDTM.

- CMENDTC: Conditional variable in character format, the predecessor is *CM.CMENDTC*.
- AENDT: Conditional variable in numeric format, conditional on whether end date is pertinent for study and is populated in SDTM.
- ASTDY: Conditional variable in numeric format, conditional on whether analysis date is pertinent for study and is populated in SDTM.

Indicator Variables for OCCDS Data Sets:

- ONTRTFL: Conditional variable in character format, indicator of whether the observation occurred while the subject was on treatment.
- PREFL: Conditional variable in character format, indicator of whether the observation occurred before the subject started treatment.
- FUPFL: Conditional variable in character format, indicator of whether the observation occurred after the subject was on treatment.

Descriptive Variables:

- AESER: Required variable in character format, the predecessor is *AE.AESER*.
- AEACN: Permissible variable in character format, the predecessor is *AE.AEACN*.
- AEREL: Permissible variable in character format, the predecessor is *AE.AEREL*.
- AEOUT: Permissible variable in character format, the predecessor is *AE.AEOUT*.
- AESCONG: Permissible variable in character format, the predecessor is *AE.AESCONG*.
- AESDISAB: Permissible variable in character format, the predecessor is *AE.AESDISAB*.
- AESDTH: Permissible variable in character format, the predecessor is *AE.AESDTH*.
- AESHOSP: Permissible variable in character format, the predecessor is *AE.AESHOSP*.
- AECONTRT: Permissible variable in character format, the predecessor is *AE.AECONTRT*.
- ARTOXGR: Permissible variable in character format, the predecessor is *AE.ARTOXGR*.

Note: There are other types of variables in the ADaM OCCDS data sets, such as "Occurrence flags", "Treatment/Dose Variables", "Standardized MedDRA Query Variables", and "Original or Prior Coding Variables". Please refer to *ADaM IG 1.2* for a more comprehensive understanding of these variables.

ADCM Specification

Table 4.5: ADCM Specification

Variable	Label	Type	Length	Controlled Terminology	Notes
STUDYID	Study Identifier	char	20		CM.STUDYID
USUBJID	Unique Subject Identifier	char	40		CM.USUBJID

(Continued)

Table 4.5: (*Continued*)

Variable	Label	Type	Length	Controlled Terminology	Notes
CMSEQ	Sequence Number	num	8		CM.CMSEQ
ASEQ	Analysis Sequence Number	num	8		Sort by USUBJID, CMTRT, ASTDT, AENDT, CMDECOD, CMSEQ. Start with 1 for the first value for the first row for each subject and increment by 1 at each successive record.
TRTP	Planned Treatment	char	40		Equal to ADSL. TRT01P
TRTA	Actual Treatment	char	40		Equal to ADSL. TRT01A
CMCAT	Category for Medication	char	40		CM.CMCAT
CMDECOD	Standardized Medication Name	char	100		CM.CMDECOD
CMSTDTC	Start Date/ Time of Medication	char	20		CM.CMSTDTC
ASTDT	Analysis Start Date	num	8		Numeric date part of CMSTDTC
ASTDTF	Analysis Start Date Imputation Flag	char	1		Equal to "D" (Day) if only the day is imputed; equal to "M" (Month) if only the month is imputed. Otherwise, is equal to NULL.
CMENDTC	End Date/ Time of Medication	char	20		CM.CMENDTC

(*Continued*)

Table 4.5: (*Continued*)

Variable	Label	Type	Length	Controlled Terminology	Notes
ASTDY	Analysis Start Relative Day	num	8		Equal to the difference between ASTDT and ADSL. TRTSDT, if ASTDT is on or after ADSL. TRTSDT, then add 1.
ADURN	Analysis Duration (N)	num	8		If CM.CMSTDTC and CM.CMENDTC are not missing, then ADURN=AENDT-ASTDT+1.
ADURU	Analysis Duration Units	char	20		Equal to "DAYS" when ADURN>1; set to "DAY" when ADURN=1; otherwise Null.
ONTRTFL	On Treatment Record Flag	char	1		Equal to "Y" if the medication was taken on or after the first dose date of study drug through the end of the treatment emergent period (ADSL. TRTSDT <= ASTDT <= ADSL.TSFENDT);
PREFL	Pre-treatment Flag	char	1		Equal to "Y" if the medication was taken before the first dose of study drug; otherwise set to null. (ASTDT<ADSL. TRTSDT)
FUPFL	Follow-Up Flag	char	1		Equal to "Y" if the AE starts on or after the end of the treatment emergent period (ASTDT >= ADSL.TSFENDT)

(*Continued*)

Table 4.5: (*Continued*)

Variable	Label	Type	Length	Controlled Terminology	Notes
AENDT	Analysis Start Date	num	8		Numeric date part of CMENDTC
ASTDTF	Analysis Start Date Imputation Flag	char	1		Equal to "D" (Day) if only the day is imputed; equal to "M" (Month) if only the month is imputed. Otherwise, is equal to NULL.
AEENDTC	End Date/ Time of Ad- verse Event	char	20		AE.AEENDTC
AENDT	Analysis Start Date	num	8		Numeric date part of AEENDTC
ASTDY	Analysis Start Relative Day	num	8		Equal to the difference between ASTDT and ADSL. TRTSDT, if ASTDT is on or after ADSL. TRTSDT, then add 1.
CMINDC	Indication	char	200		CM.CMINDC
CMROUTE	Route of Administration	char	40		CM.CMROUTE
CMENRTPT	End Relative to Reference Time Point	char	20		CM.CMENRTPT

ADCM Programming

In order to create the ADCM domain, we start by merging SDTM.CM and ADAM.ADSL. Please see code below.

Note: For every domain, the corresponding SAS name is domain.sas. For example, ADCM is created by adcm.sas, so all code in this section belongs to adcm.sas.

```
/*Begin writing SAS program - merge SDTM.CM and ADAM.ADSL.sas*/
data ADCM1;
  merge SDTM.CM(in=a) ADAM.ADSL(in=b drop=STUDYID);        ❶
  by USUBJID;
```

```
  if a and b;
run;
```

❶ Merge CM_SUPPCM and ADAM.ADSL.

```
data ADCM2;
  set ADCM1;
  length ADURU $20 ONTRTFL PREFL ASTDTF $1 ASTDT AENDT ADURN 8. ;
  /*Derive ASTDT, AENDT */

  ASTDT=input(substr(CMSTDTC,1,10),yymmdd10.);
  AENDT=input(substr(CMENDTC,1,10),yymmdd10.);
  format ASTDT AENDT yymmdd10.;
  /*Derive ONTRTFL, PREFL, FUPFL, EMDESC*/

  if TRTSDT <= ASTDT <= TRTEDT then ONTRTFL = "Y";
  else if . < ASTDT < TRTSDT then PREFL = "Y";
  else if ASTDT > TRTEDT and TRTEDT^=. then  FUPFL = "Y";
  /*Derive ASTDY*/

  if nmiss(ASTDT,TRTSDT)=0 then ASTDY=ASTDT-TRTSDT+(ASTDT>=TRTSDT);
  /*Derive ASTDTF*/
  if length(CMSTDTC) >= 10 then ASTDTF="";
  else if length(CMSTDTC)=7 then ASTDTF="D";
  else if length(CMSTDTC)=4 then ASTDTF="M";
  /*Derive ADURN, ADURU*/
  if cmiss(CMSTDTC,CMENDTC)=0 then ADURN=AENDT-ASTDT+1;
  if ADURN>1 then ADURU="DAYS";
  else if ADURN=1 then ADURU="DAY";
  else ADURU="";
  /*Derive TRTP, TRTA*/

  TRTA=TRT01A;
  TRTP=TRT01P;
run;
proc sort data=ADCM2;
  by USUBJID CMTRT ASTDT AENDT CMDECOD CMSEQ;
run;
/*Derive ASEQ*/

data Final;
  set ADCM2;
  by USUBJID CMTRT ASTDT AENDT CMDECOD CMSEQ;
  if first.USUBJID then ASEQ = 0;
     ASEQ+1;
  output;
run;
```

❶ Create variable ASEQ variable. This variable is used to uniquely record analysis sequence number of subjects within a domain.

```
libname ADAM ".../directory";
data ADAM.ADCM(label="Concomitant Medications Analysis Data Set");
```

```
/*Assign variable attributes such as label and length to conform with
ADAM.ADSL Specification (these will also be the same attributes as the ADAM
IG).*/

attrib
 STUDYID  label = "Study Identifier"                              length = $20
 USUBJID  label = "Unique Subject Identifier"                     length = $40
 CMSEQ    label = "Sequence Number"                               length = 8
 ASEQ     label = "Analysis Sequence Number"                      length = 8
 CMDECOD  label = "Dictionary-Derived Term"                       length = $200
 ASTDT    label = "Analysis Start Date"                           length = 8
 ASTDTF   label = "Analysis Start Date Imputation Flag"           length = $1
 ONTRTFL  label = "On Treatment Record Flag"                      length = $1
 PREFL    label = "Pre-treatment Flag"                            length = $1
 FUPFL    label = "Follow-Up Flag"                                length = $1
 AENDT    label = "Analysis Start Date"                           length = 8
 ASTDY    label = "Analysis Start Relative Day"                   length = 8
 CMTRT    label = "Reported name of drug, Medication or Therapy"  length = $200
 CMINDC   label = "Indication"                                    length = $200
 CMROUTE  label = "Route of Administration"                       length = $40
 CMSTDTC  label = "Start Date/Time of Medication"                 length = $20
 CMENDTC  label = "End Date/Time of Medication"                   length = $20
 CMENRTPT label = "End Relative to Reference Time Point"          length = $20
     ;
   set FINAL;
   keep STUDYID USUBJID CMSEQ CMDECOD ASTDT ASTDTF ONTRTFL PREFL FUPFL AENDT
ASTDY CMTRT CMINDC CMROUTE CMSTDTC CMENDTC CMENRTPT
     ;
run;
```

Figure 4.6: ADAM.ADCM

	STUDYID	USUBJID	CMSEQ	CMDECOD	ASTDT	ASTDTF	ONTRTFL	PREFL	FUPFL
1	ABC-001	ABC-001-001-001	1	PRUNELLA	2021-01-06				Y
2	ABC-001	ABC-001-001-002	1	PRUNELLA	2021-01-06				Y
3	ABC-001	ABC-001-001-006	1	PRUNELLA	2021-01-07				Y
4	ABC-001	ABC-001-001-009	1	PRUNELLA	2021-01-09				Y

AENDT	ASTDY	CMTRT	CMINDC	CMROUTE	CMSTDTC
2021-01-06	5	PRUNE JUICE	CONSTIPATION	ORAL	2021-01-06
2021-01-06	5	PRUNE JUICE	CONSTIPATION	ORAL	2021-01-06
2021-01-07	6	PRUNE JUICE	CONSTIPATION	ORAL	2021-01-07
2021-01-09	8	ASPIRIN	FEVER	ORAL	2021-01-09

CMENDTC	CMENRTPT
2021-01-06	
2021-01-06	
2021-01-07	
2021-01-09	

Chapter 5: Case Report Tabulation Data Definition (Define-XML)

The Case Report Tabulation Data Definition Specification (Define-XML) transmits metadata that describes any tabular data set structure, such as the SDTM and ADaM data that we have been describing throughout this book. The FDA and PMDA requires Define-XML in addition to, and to describe in more detail, the CDISC SDTM and ADaM and data for drug submission as it improves the reviewer's understanding of the entire data submission package. This chapter describes the process of generating a Define-XML file.

The following section describes some general concepts of XML code. We do not cover details of XML coding as the Define-XML format continues to evolve and can be created in a plethora of ways, including SAS programming, off the shelf software, etc. The purpose of this section is to give the reader an overview of why we create Define-XML.

5.1 Structure of Define-XML

The Define-XML file forms a tree structure that starts from a root to branches to leaves and upward. The basic syntax for the XML language is from the start element's tag to the end element's tag. For example, for the tree shown below, type is an element. Name, color, and height are all features of the element.

```
<tree>
<type= "deciduous tree">
  <name>maple<name>
  <color>orange<color>
  <height>10 meters<height>
</type>
<tree>
```

The basic syntax of XML is shown below.

```
'<?xml version="1.0" encoding="UTF-8" ?>' /                                    ❶
'<ODM' /                                                                        ❷
    'xmlns="http://www.cdisc.org/ns/odm/v1.3"' /
    'xmlns:xlink="http://www.w3.org/1999/xlink"' /
    'xmlns:def="http://www.cdisc.org/ns/def/v2.0"' /
    'ODMVersion="1.3.2"' /
    'File="ABC-001-Define-XML_2.0.0"' /
    'Originator="CDISC XML Technologies Team"' /
    'SourceSystem="MH-System">' /
    'SourceSystemVersion="2.0.1">' /
        '<GlobalVariables>' /                                                   ❸
            '<StudyName>' ABC-001 '</StudyName>' /
            '<StudyDescription>' Pain Test Study'</StudyDescription>' /
            '<ProtocolName>' ABC-001 '</ProtocolName>' /
        '</GlobalVariables>' /
        '<MetaDataVersion OID="CDISC.SDTM.1 " /
            'Name="ABC-001 Pain Test " ' /
            'Description="ABC-001 Pain Test " /
            'def:DefineVersion="2.0.0"' /
            'def:StandardName="' CDISC SDTM"' /
            'def:StandardVersion="1">' /

        '<ItemGroupDef OID="DM"' /                                              ❹
            'Domain="DM"' /
            'Name=" DM "' /
            'Repeating="Yes "' /
            'IsReferenceData="No"' /
            'SASDatasetName="DM "' /
            'Purpose="Tabulation "' /
            'def:Structure="One Record per Subject "' /
            'def:Class="SPECIAL PURPOSE"' /
            'def:DomianKeys="USUBJID"' /
            'def:ArchiveLocationID="Location.DM '">' /;


        '<ItemRef ItemOID="STUDYID "' /                                         ❺
            'OrderNumber="1 "' /
            'Mandatory=" Yes "' /;


        '<def:leaf ID="Location.DM '" xlink:href="DM.xpt">' /                   ❻
            '<def:title>' DM.xpt '</def:title>' /
        '</def:leaf>' /
    '</ItemGroupDef>';
```

Some notes about XML code style meaning are described below.

❶   Every XML program starts with the XML declaration, or the processing instruction indicating the document is in XML format.

❷   The <ODM> statement is the root element. The following "xmls" are Namespaces, which are mechanisms by which element and attribute names are assigned to groups. Correspondingly, "ODMVersion", "File", "Originator", "SourceSystem", "SourceSystemVersion" are the attributes of <ODM>.

❸   The following are elements: <GlobalVariables>, <StudyName>, <StudyDescription>, <ProtocolName>, < MetaDataVersion>, < ItemGroupDef >, < ItemRef >. Elements are the basic compositions of XML file, splitting a document into separate regions with a hierarchy.

❹   The element <ItemGroupDef> is used to define the domain level metadata, such as DM and DS in SDTM and ADSL and ADAE in ADaM.

❺   The element <ItemRef> is used to define the variable level metadata.

❻   The element <leaf> is used to link the external documents such as annotation aCRF, XX.XPT file, where 'XX' is the SDTM or ADaM domain, etc.

Note: Elements should be properly opened and closed. For example, if the open element is <ItemGroupDef>, then we also need closed element <ItemGroupDef>.

5.2 The Process of Creating Define-XML

In order to generate Define-XML, there are four steps:

1. Create the metadata and use SAS to create Define-XML components.

   ○ For the purpose of this book and to help readers visually understand what we mean by metadata, we demonstrate by populating the metadata in a spreadsheet.

2. Create XPT files using SAS.
3. Link for external documents such as aCRF and XPT files.
4. Construct Define.XML.

5.2.1 Create Metadata Spreadsheet and Create Define-XML Components

It is essential to create the metadata spreadsheet for each of the sections consistently. The metadata has five main parts including: header metadata, table of contents metadata, variable level metadata, controlled terminology metadata, and computation method metadata.

Figure 5.1 shows the header metadata, which includes the information of study, protocol, SDTM (also created for ADaM) version, and the version of style sheet.

The following process shows how to read header metadata from the spreadsheet one at a time and then write it into the XML as a TXT file.

1. Read the "header metadata" tab into SAS from the "define.xml" spreadsheet.
2. The meta_header.txt has the information about the header metadata that will populate the defile.xml elements including file name, study name, study description, protocol name, standard, version, and style sheet.
3. The last three lines populate the link to the aCRF document.

Figure 5.1: Header Metadata

FILE	STUDY	STUDY_DESCRIPTION	PROTOCOL_NAME	STANDARD	VERSION	STYLESHEET
ABC-001	ABC-001	A Phase I, Open Label, To Evaluate the Efficacy of Drug A in Subjects Who Suffer with Pain	ABC-001	SDTM	1	define2-0-0.xsl

The following process shows how to read in the table of contents metadata from the spreadsheet individually (can be done by using different tabs for each domain or other ways) and then write it into XML as a TXT file.

1. Read the "TOC metadata" tab into SAS from the "define.xml" XLSX sheet.
2. The meta_TOC.txt contains the information for the domain level metadata that will be populated in the defile.xml elements, including domain name, purpose, structure, location, key variables, etc.

Figure 5.2 shows the table of contents metadata, which lists all the data sets included for the drug submission.

Figure 5.2: Table of Contents Metadata

DATASET	REPEATING	REFERENCE_DATA	LABEL	CLASS	STRUCTURE	PURPOSE	KEYS	LOCATION	ORDER
DM	Yes	No	Demographics	SPECIAL PURPOSE	One Record per Subject	Tabulation	STUDYID, DOMAIN, USUBJID, SUBJID, RFSTDTC, RFENDTC	DM.xpt	1
EX	Yes	No	Exposure	INTERVENTIONS	One Record per Protocol-Specified study treatment, constant-dosting invertal, per subject	Tabulation	STUDYID, DOMAIN, USUBJID, EXSEQ, EXSTDTC, EXENDTC, EXDOSE	EX.xpt	2
...	...	...	...	...	...	...	...	...	...

The following process shows how to read variable level metadata from the spreadsheet individually (can be done by using different tabs for each domain or other ways) and then write it into the XML as a TXT file.

1. Read the "variable metadata" tab into SAS from the "define.xml" XLSX sheet.
2. The meta_Variable.txt has the information about the variable level metadata that will be populated in the define.xml elements including variable name, variable type, variable length, variable label, comments, etc., which are columns in the spreadsheet in Figure 5.3.

Note: The variable level metadata is used to describe the variables within each domain. The *KEY_SEQUENCE* variable is used to provide a sequential number of key variables so that the style sheet can use it to populate the *KEYS* in the table of contents metadata section in the define file automatically.

Figure 5.3 shows the variable level metadata, which describes each variable within each data set.

Figure 5.3: Variable Level Metadata

DATASET	VARIABLE	LABEL	TYPE	LENGTH	CONTROLLED TERMINOLOGY	ORIGIN	COMMENTS	KEY_SEQUENCE
DM	STUDYID	Study Identifier	text	20		Protocol	Set to "ABC-001"	1
DM	DOMAIN	Domain Abbreviation	text	2	DOMAIN	Assigned	Set to "DM"	2
DM	USUBJID	Unique Subject Identifier	text	20		Derived	Concatenate among three varaibles " STUDYID-SITED-SUBJID"	3
DM	SUBJID	Subject Identifier for the Study	text	20		CRF		4
DM	RFSTDTC	Subject Reference Start Date/Time	text	20		Derived	Equal to RFXSTDTC	5
DM	RFENDTC	Subject Reference End Date/Time	text	20		Derived	Equal to RFXENDTC	6
...	...	...		...	...	...	...	...

The following process shows how to read controlled terminology metadata from the spreadsheet one at a time and then write it into the XML as a TXT file.

1. Read the "Controlled Terminology" tab into SAS from the "define.xml" XLSX sheet.
2. The Meta_Codelist.txt has the information about the controlled terminology metadata that will be populated in the defile.xml elements, including codelist name, codelist type, codelist value, decode value, etc.

Note: The controlled terminology metadata is used to display the coded terms under each variable within each domain when *DECODE* is "Yes". The *RANK* variable is used to provide a sequential number of code value.

Figure 5.4 shows the controlled terminology metadata, which is used to associate the variables into categories.

Figure 5.4: Controlled Terminology Metadata

CODELIST_NAME	CODELIST_VALUE	CODE_VALUE	TRANSLATED	TYPE	DECODE	RANK
DM	Domain Abbreviation (DM)	DM	Demographics	text	YES	
RACE		WHITE	White		White	1
RACE		BLACK OR AFRICAN AMERICAN	Black or African American		Black or African American	2
RACE		ASIAN	Asian		Asian	3
RACE		AMERICAN INDIAN OR ALASKA NATIVE	American Indian or Alaska Native		American Indian or Alaska Native	4
RACE		NATIVE HAWAIIAN OR OTHER PACIFIC ISLANDER	Native Hawaiian or Other Pacific Islander		Native Hawaiian or Other Pacific Islander	5
RACE		NOT REPORTED				6
RACE		MULTIPLE	Multiple		Multiple	7
RACE		OTHER	Other		Other	8

The following process shows how to read computational method metadata from the spreadsheet one at a time and then write it into the XML as a TXT file.

1. Read the "Computational Method" tab into SAS from the "define.xml" XLSX sheet.
2. The meta_Computation.txt has the information about the computational methods metadata that will be populated in the define.xml elements, including computation name, computation method, type, description.

Figure 5.5: Computation Methods Metadata

COMPUTATION_NAME	COMPUTATION_METHOD	TYPE	DESCRIPTION
DM.USUBJID	Algorithm to derive DM.USUBJID	Computation	Concatenate among three variables: STUDYID-SITEID-SUBJID
DM.RFSTDTC	Algorithm to derive RFSTDTC	Computation	Date and time of the first dosing for every subject
RFENDTC	Algorithm to derive RFENDTC	Computation	Date and time of the last dosing for every subject

5.2.2 Create XPT Files

There are at least two ways to create XPT files using SAS. The first way is to use a DATA step. The second way is to use PROC COPY. The syntax for both ways is straightforward as shown in the code below but notice when creating a pathway for storing the XPT file that you must specify the name of the XPT file into the XPORT engine, e.g., <"Pathway\.xpt">.

```
libname sasfile <"Pathway1">;
libname xptfile XPORT <"Pathway\.xpt">;
```

```
data xptfile.ts;
  set sasfile.<domain name>;
run;

proc copy in=sasfile
     out=xptfile memtype=data;
select <sheet name>;
run;
```

5.2.3 Link for External Documents

Use def:leaf to link the external documents, such as the annotation CRF Form, XPT files, etc. The Define.XML would have a detailed process for this section.

5.2.4 Construct Define.XML

Convert all TXT files into XML files to be viewed in an internet browser.

Conclusion

This book contains a lot of practical CDISC (SDTM and ADaM) domain examples created using the SAS programming language. We have explained basic pharmaceutical industry knowledge (Chapter 1), how to annotate the CRF (Chapter 2), how to create SDTM Specifications and program SDTM domains (Chapter 3), how to create ADaM Specifications and program ADaM data sets (Chapter 4), and lastly, we briefly explained Define-XML (Chapter 5). After reading this book, readers will be able to understand standardized clinical data, as well as how to create it, in the pharmaceutical industry.

Learn more about the authors and this book by visiting http://support.sas.com/case to access the example code and data, read the latest reviews, get updates, and more.

Appendix

A.1 Raw Data Spreadsheet

RAW.TS

	A	B	C	D	E	F	G	H
1	STUDYID	DOMAIN	TSSEQ	TSPARMCD	TSPARM	TSVAL	TSVALNF	TSVCDREF
2	ABC-001	TS	1	ACTSUB	Actual Number of Subjects	10		
3	ABC-001	TS	1	ADAPT	Adaptive Design	N		CDISC
4	ABC-001	TS	1	ADDON	Added on to Existing Treatments	N		CDISC
5	ABC-001	TS	1	AGEMAX	Planned Maximum Age of Subjects	P50Y		ISO 8601
6	ABC-001	TS	1	AGEMIN	Planned Minimum Age of Subjects	P18Y		ISO 8601
7	ABC-001	TS	1	DCUTDESC	Data Cutoff Description	Final Database Lock		
8	ABC-001	TS	1	DCUTDTC	Data Cutoff Date	2021-01-21		ISO 8601
9	ABC-001	TS	1	DOSE	Dose per Administration			
10	ABC-001	TS	1	DOSFRQ	Dosing Frequency	UNKNOWN		CDISC
11	ABC-001	TS	1	FCNTRY	Planned Country of Investigational Sites	USA		ISO 3166
12	ABC-001	TS	1	HLTSUBJI	Healthy Subject Indicator	N		CDISC
13	ABC-001	TS	1	INDIC	Trial Disease/Condition Indication	Acute pain		SNOMED
14	ABC-001	TS	1	INTMODEL	Intervention Model	PARALLEL		CDISC
15	ABC-001	TS	1	INTTYPE	Intervention Type	SURGERY		CDISC
16	ABC-001	TS	1	LENGTH	Trial Length	P10D		ISO 8601
17	ABC-001	TS	1	NARMS	Planned Number of Arms	1		
18	ABC-001	TS	1	OBJPRIM	Trial Primary Objective	To evaluate the efficacy of Drug A in subjects who suffer with Pain		

19	ABC-001	TS	1	OBJSEC	Trial Secondary Objective	To evaluate the safety of Drug A in subjects who suffer with Pain		
20	ABC-001	TS	1	OUTMSPRI	Primary Outcome Measure	SEIZURE FREQUENCY		
21	ABC-001	TS	1	OUTMSSEC	Secondary Outcome Measure	SEIZURE FREQUENCY		
22	ABC-001	TS	1	OUTMSADD	Additional Outcome Measure	SEIZURE FREQUENCY		
23	ABC-001	TS	1	PCLAS	Pharmacologic Class		NAV	ISO 21090
24	ABC-001	TS	1	PLANSUB	Planned Number of Subjects	10		
25	ABC-001	TS	1	RANDQT	Randomization Quotient	N		
26	ABC-001	TS	1	RANDOM	Trial is Randomized	N		CDISC
27	ABC-001	TS	1	REGID	Registry Identifier	NCTI123456789		ClinicalTrials.GOV
28	ABC-001	TS	1	SENDTC	Study End Date	2021-01-20		EudraCT
29	ABC-001	TS	1	SEVCRIT	Severity Criteria	NA		ISO 8601
30	ABC-001	TS	1	SEXPOP	Sex of Participants	BOTH		
31	ABC-001	TS	1	SPONSOR	Clinical Study Sponsor	MALE		CDISC
32	ABC-001	TS	1	SSTDTC	Study Start Date	XX Pharma		
33	ABC-001	TS	1	STOPRULE	Study Stop Rules	NA		ISO 8601
34	ABC-001	TS	1	STRATFCT	Stratification Factor	NONE		
35	ABC-001	TS	1	STYPE	Study Type	PAIN		
36	ABC-001	TS	1	TBLIND	Trial Blinding Schema	OBSERVATIONAL		CDISC
37	ABC-001	TS	1	TCNTRL	Control Type	OPEN LABEL		CDISC
38	ABC-001	TS	1	TINDTP	Trial Intent Type	NA		SNOMED
39	ABC-001	TS	1	TITLE	Trial Title	TREATMENT		CDISC
40	ABC-001	TS	1	TPHASE	Trial Phase Classification	PHASE I TRIAL		
41	ABC-001	TS	1	TRT	Investigational Therapy or Treatment	PHASE I TRIAL		CDISC
42	ABC-001	TS	1	TTYPE	Trial Type	TREATMENT		

RAW.TA

	A	B	C	D	E	F	G	H	I	J
1	STUDYID	DOMAIN	ARMCD	ARM	TAETORD	ETCD	ELEMENT	TABRANCH	TATRANS	EPOCH
2	ABC-001	TA	DRUG A	DRUG A 10 MG	1	SCRE	Screening			SCREENING
3	ABC-001	TA	DRUG A	DRUG A 10 MG	2	PRE	Pretreatment			PRETREATMENT
4	ABC-001	TA	DRUG A	DRUG A 10 MG	3	TRT	Treatment		if diease progression, then go to Satey Follow Up Epoch	OPEN LABEL TREATMENT
5	ABC-001	TA	DRUG A	DRUG A 10 MG	4	SFU	Safety Follow Up			SAFETY FOLLOW UP

RAW.TE

	A	B	C	D	E	F	G
1	STUDYID	DOMAIN	ETCD	ELEMENT	TESTRL	TEENRL	TEDUR
2	ABC-001	TE	SCRE	Screening	Inform consent obtained	Admission to clinic	
3	ABC-001	TE	PRE	Pretreatment	Admission to clinic	First dose of study drug, where drug is Drug A	
4	ABC-001	TE	TRT	Treatment	First dose of study drug, where drug is Drug A	First dose of study drug on Day 3	P3D
5	ABC-001	TE	SFU	Safety Follow Up	48 hours after last dose	At Trial Exit	

RAW.TV

	A	B	C	D	E	F	G
1	STUDYID	DOMAIN	VISITNUM	VISIT	ARMCD	ARM	TVSTRL
2	ABC-001	TV		1 SCREENING	DRUG A	DRUG A 10 mg	Inform Consent Signed
3	ABC-001	TV		2 DAY -1	DRUG A	DRUG A 10 mg	Admission to Clinic
4	ABC-001	TV		3 DAY 1	DRUG A	DRUG A 10 mg	Start of Day 1
5	ABC-001	TV		4 DAY 2	DRUG A	DRUG A 10 mg	Start of Day 2
6	ABC-001	TV		5 DAY 3	DRUG A	DRUG A 10 mg	Start of Day 3
7	ABC-001	TV		6 FOLLOW UP	DRUG A	DRUG A 10 mg	10 Days after last dose

RAW.TI

	A	B	C	D	E
1	STUDYID	DOMAIN	IETESTCD	IETEST	IECAT
2	ABC-001	TI	IN01	Subject will sign the inform consent form	INCLUSION
3	ABC-001	TI	IN02	Subjects are 18 years to 50 years of age	INCLUSION
4	ABC-001	TI	IN03	Female subjects are no child-bearing potential	INCLUSION
5	ABC-001	TI	EX01	Positive for HIV1 or HIV2 antibody during screening	EXCLUSION
6	ABC-001	TI	EX02	Cigaratte smoking during the past 3 monthes at screening	EXCLUSION
7	ABC-001	TI	EX03	History of significant alcohol consumption in the past 3 months before screening	EXCLUSION

RAW.DM

	E	F	G	H	I	J	K	L	M
1	FORM	BRTHDAT	AGE	AGEU	SEX	ETHNIC	RACE_WHITE	RACE_HAWAIIAN	RACE_BLACK
2	Demographics	2/3/2000		20 YEARS	Female	Not Hispanic or Latino	WHITE		
3	Demographics	4/6/2000		20 YEARS	Female	Not Hispanic or Latino	WHITE		
4	Demographics	8/11/1990		30 YEARS	Male	Not Hispanic or Latino	WHITE		
5	Demographics	9/9/1985		35 YEARS	Male	Not Hispanic or Latino	WHITE		
6	Demographics	2/3/1998		22 YEARS	Male	Not Hispanic or Latino	WHITE		
7	Demographics	8/9/1986		34 YEARS	Male	Not Hispanic or Latino	WHITE		
8	Demographics	2/1/1996		24 YEARS	Male	Not Hispanic or Latino	WHITE		
9	Demographics	9/5/2000		20 YEARS	Male	Not Hispanic or Latino			Black or African American
10	Demographics	7/8/1990		30 YEARS	Male	Not Hispanic or Latino			Black or African American
11	Demographics	8/20/1990		30 YEARS	Male	Not Hispanic or Latino			Black or African American

N	O		P	Q	R
RACE_ASIAN	RACE_AINDIAN		RACE_NOREPORT	RACE_UNKNOWN	RACE_OTHER
	AMERICAN INDIAN OR ALASKA NATIVE				

RAW.DS

	A	B	C	D	E	F	G	H	I
1	STUDYID	COUNTRY	SITE	SUBJID	EVENT	DSSTDAT	FORM	DSDECOD	DSDISC
2	ABC-001	United States	001	001	Termination		1/4/2021 End of Dosing-drug A	COMPLETED	
3	ABC-001	United States	001	002	Termination		1/4/2021 End of Dosing-drug A	COMPLETED	
4	ABC-001	United States	001	003	Termination		1/4/2021 End of Dosing-drug A	COMPLETED	
5	ABC-001	United States	001	004	Termination		1/4/2021 End of Dosing-drug A	COMPLETED	
6	ABC-001	United States	001	005	Termination		1/4/2021 End of Dosing-drug A	COMPLETED	
7	ABC-001	United States	001	006	Termination		1/4/2021 End of Dosing-drug A	COMPLETED	
8	ABC-001	United States	001	007	Termination		1/4/2021 End of Dosing-drug A	COMPLETED	
9	ABC-001	United States	001	008	Termination		1/4/2021 End of Dosing-drug A	COMPLETED	
10	ABC-001	United States	001	009	Termination		1/4/2021 End of Dosing-drug A	COMPLETED	
11	ABC-001	United States	001	010	Termination		1/4/2021 End of Dosing-drug A	COMPLETED	

RAW.AE

	A	B	C	D	E	F	G	H	I	J	K	L	M	N	O
1	STUDYID	COUNTRY	SITEID	SUBJID	ESEQ	FORM	FSEQ	IGSEQ	AEYN	AETERM	AESER	AESMIE	AESHOSP	AESDTH	AESLIFE
2	ABC-001	United States	001	001		1 Adverse Events	1	1	Y	Headache	No	No	No	No	No
3	ABC-001	United States	001	002		1 Adverse Events	1	1	Y	Sore Neck	No	No	No	No	No
4	ABC-001	United States	001	003		1 Adverse Events	1	1	Y	Sore Neck	No	No	No	No	No
5	ABC-001	United States	001	004		1 Adverse Events	1	1	Y	Sore Neck	No	No	No	No	No
6	ABC-001	United States	001	005		1 Adverse Events	1	1	Y	Sore Neck	No	No	No	No	No
7	ABC-001	United States	001	006		1 Adverse Events	1	1	Y	Sore Neck	No	No	No	No	No
8	ABC-001	United States	001	007		1 Adverse Events	1	1	Y	Sore Neck	No	No	No	No	No
9	ABC-001	United States	001	008		1 Adverse Events	1	1	Y	Sore Neck	No	No	No	No	No
10	ABC-001	United States	001	009		1 Adverse Events	1	1	Y	Sore Neck	No	No	No	No	No
11	ABC-001	United States	001	010		1 Adverse Events	1	1	Y	Headache	No	No	No	No	No

P	Q	R	S	T	U	V	W	X	Y
AESDISAB	AESCONG	AESDAT	AEENDAT	AEACN	AEREL	AESEV	AEOUT	SOC	PT
No	No	1/5/2021	1/5/2021	Drug Not Changed	No	Mild	Recoverd or Resolved	Nervous system disorders	Headaches
No	No	1/7/2021	1/7/2021	Drug Not Changed	NO	Mild	Recoverd or Resolved	Musculoskeletal and connective tissue disorders	Neck Pain
No	No	1/7/2021	1/7/2021	Drug Not Changed	No	Moderate	Recoverd or Resolved	Musculoskeletal and connective tissue disorders	Neck Pain
No	No	1/7/2021	1/7/2021	Drug Not Changed	NO	Moderate	Recoverd or Resolved	Musculoskeletal and connective tissue disorders	Neck Pain
No	No	1/7/2021	1/7/2021	Drug Not Changed	No	Moderate	Recoverd or Resolved	Musculoskeletal and connective tissue disorders	Neck Pain
No	No	1/7/2021	1/7/2021	Drug Not Changed	No	Moderate	Recoverd or Resolved	Musculoskeletal and connective tissue disorders	Neck Pain
No	No	1/7/2021	1/7/2021	Drug Not Changed	NO	Moderate	Recoverd or Resolved	Musculoskeletal and connective tissue disorders	Neck Pain
No	No	1/7/2021	1/7/2021	Drug Not Changed	No	Mild	Recoverd or Resolved	Musculoskeletal and connective tissue disorders	Neck Pain
No	No	1/9/2021	1/9/2021	Drug Not Changed	NO	Mild	Recoverd or Resolved	Musculoskeletal and connective tissue disorders	Neck Pain
No	No	1/9/2021	1/9/2021	Drug Not Changed	Yes	Mild	Recoverd or Resolved	Nervous system disorders	Headaches

Z	AA	AB	AC	AD	AE	AF	AG	AH
LLT	LLTCD	HLT	HLTCD	PT_CD	HLGT	HLGTCD	DICTYPE	MEDDRUG
Headache	10019211	Headaches NEC	10019233	10019211	Headaches	10019231	MedDRA	MedDRA 20.1
Neck Pain	10028836	Musculoskeletal and connective tissue disorders	10068757	10028836	Musculoskeletal and connective tissue disorders NEC	10028393	MedDRA	MedDRA 20.1
Neck Pain	10028836	Musculoskeletal and connective tissue disorders	10068757	10028836	Musculoskeletal and connective tissue disorders NEC	10028393	MedDRA	MedDRA 20.1
Neck Pain	10028836	Musculoskeletal and connective tissue disorders	10068757	10028836	Musculoskeletal and connective tissue disorders NEC	10028393	MedDRA	MedDRA 20.1
Neck Pain	10028836	Musculoskeletal and connective tissue disorders	10068757	10028836	Musculoskeletal and connective tissue disorders NEC	10028393	MedDRA	MedDRA 20.1
Neck Pain	10028836	Musculoskeletal and connective tissue disorders	10068757	10028836	Musculoskeletal and connective tissue disorders NEC	10028393	MedDRA	MedDRA 20.1
Neck Pain	10028836	Musculoskeletal and connective tissue disorders	10068757	10028836	Musculoskeletal and connective tissue disorders NEC	10028393	MedDRA	MedDRA 20.1
Neck Pain	10028836	Musculoskeletal and connective tissue disorders	10068757	10028836	Musculoskeletal and connective tissue disorders NEC	10028393	MedDRA	MedDRA 20.1
Neck Pain	10028836	Musculoskeletal and connective tissue disorders	10068757	10028836	Musculoskeletal and connective tissue disorders NEC	10028393	MedDRA	MedDRA 20.1
Headache	10019211	Headaches NEC	10019233	10019211	Headaches	10019231	MedDRA	MedDRA 20.1

RAW.EX

	A	B	C	D	E	F	G
1	STUDYID	COUNTRY	SITEID	SUBJID	EVENT	EVENT_DT	ESEQ
2	ABC-001	United States	001	001	drug A treatment 1 day 1	1/2/2021	1
3	ABC-001	United States	001	001	drug A treatment 1 day 2	1/3/2021	1
4	ABC-001	United States	001	001	drug A treatment 1 day 3	1/4/2021	1
5	ABC-001	United States	001	002	drug A treatment 1 day 1	1/2/2021	1
6	ABC-001	United States	001	002	drug A treatment 1 day 2	1/3/2021	1
7	ABC-001	United States	001	002	drug A treatment 1 day 3	1/4/2021	1
8	ABC-001	United States	001	003	drug A treatment 1 day 1	1/2/2021	1
9	ABC-001	United States	001	003	drug A treatment 1 day 2	1/3/2021	1
10	ABC-001	United States	001	003	drug A treatment 1 day 3	1/4/2021	1
11	ABC-001	United States	001	004	drug A treatment 1 day 1	1/2/2021	1
12	ABC-001	United States	001	004	drug A treatment 1 day 2	1/3/2021	1
13	ABC-001	United States	001	004	drug A treatment 1 day 3	1/4/2021	1
14	ABC-001	United States	001	005	drug A treatment 1 day 1	1/2/2021	1
15	ABC-001	United States	001	005	drug A treatment 1 day 2	1/3/2021	1
16	ABC-001	United States	001	005	drug A treatment 1 day 3	1/4/2021	1
17	ABC-001	United States	001	006	drug A treatment 1 day 1	1/2/2021	1
18	ABC-001	United States	001	006	drug A treatment 1 day 2	1/3/2021	1
19	ABC-001	United States	001	006	drug A treatment 1 day 3	1/4/2021	1
20	ABC-001	United States	001	007	drug A treatment 1 day 1	1/2/2021	1
21	ABC-001	United States	001	007	drug A treatment 1 day 2	1/3/2021	1
22	ABC-001	United States	001	007	drug A treatment 1 day 3	1/4/2021	1
23	ABC-001	United States	001	008	drug A treatment 1 day 1	1/2/2021	1
24	ABC-001	United States	001	008	drug A treatment 1 day 2	1/3/2021	1
25	ABC-001	United States	001	008	drug A treatment 1 day 3	1/4/2021	1
26	ABC-001	United States	001	009	drug A treatment 1 day 1	1/2/2021	1
27	ABC-001	United States	001	009	drug A treatment 1 day 2	1/3/2021	1
28	ABC-001	United States	001	009	drug A treatment 1 day 3	1/4/2021	1
29	ABC-001	United States	001	010	drug A treatment 1 day 1	1/2/2021	1
30	ABC-001	United States	001	010	drug A treatment 1 day 2	1/3/2021	1
31	ABC-001	United States	001	010	drug A treatment 1 day 3	1/4/2021	1

H	I	J	K	L	M	N	O	P	Q	R
EXREFID	FORM	FSEQ	IGSEQ	EXSTDAT	EXENDAT	EXDOSE	EXDOSU	EXDOSTXT	EXDOSFRQ	EXROUTE
1	Drug Administration-drug A	1	1	1/2/2021 10:00	1/2/2021 10:00	10	mg		Daily	Oral
1	Drug Administration-drug A	1	1	1/3/2021 10:00	1/3/2021 10:00	10	mg		Daily	Oral
1	Drug Administration-drug A	1	1	1/4/2021 10:00	1/4/2021 10:00	10	mg		Daily	Oral
1	Drug Administration-drug A	1	1	1/2/2021 10:00	1/2/2021 10:00	10	mg		Daily	Oral
1	Drug Administration-drug A	1	1	1/3/2021 10:00	1/3/2021 10:00	10	mg		Daily	Oral
1	Drug Administration-drug A	1	1	1/4/2021 10:00	1/4/2021 10:00	10	mg		Daily	Oral
1	Drug Administration-drug A	1	1	1/2/2021 10:00	1/2/2021 10:00	10	mg		Daily	Oral
1	Drug Administration-drug A	1	1	1/3/2021 10:00	1/3/2021 10:00	10	mg		Daily	Oral
1	Drug Administration-drug A	1	1	1/4/2021 10:00	1/4/2021 10:00	10	mg		Daily	Oral
1	Drug Administration-drug A	1	1	1/2/2021 10:00	1/2/2021 10:00	10	mg		Daily	Oral

1	Drug Administration-drug A	1	1	1/3/2021 10:00	1/3/2021 10:00	10 mg	Daily	Oral
1	Drug Administration-drug A	1	1	1/4/2021 10:00	1/4/2021 10:00	10 mg	Daily	Oral
1	Drug Administration-drug A	1	1	1/2/2021 10:00	1/2/2021 10:00	10 mg	Daily	Oral
1	Drug Administration-drug A	1	1	1/3/2021 10:00	1/3/2021 10:00	10 mg	Daily	Oral
1	Drug Administration-drug A	1	1	1/4/2021 10:00	1/4/2021 10:00	10 mg	Daily	Oral
1	Drug Administration-drug A	1	1	1/2/2021 10:00	1/2/2021 10:00	10 mg	Daily	Oral
1	Drug Administration-drug A	1	1	1/3/2021 10:00	1/3/2021 10:00	10 mg	Daily	Oral
1	Drug Administration-drug A	1	1	1/4/2021 10:00	1/4/2021 10:00	10 mg	Daily	Oral
1	Drug Administration-drug A	1	1	1/2/2021 10:00	1/2/2021 10:00	10 mg	Daily	Oral
1	Drug Administration-drug A	1	1	1/3/2021 10:00	1/3/2021 10:00	10 mg	Daily	Oral
1	Drug Administration-drug A	1	1	1/4/2021 10:00	1/4/2021 10:00	10 mg	Daily	Oral
1	Drug Administration-drug A	1	1	1/2/2021 10:00	1/2/2021 10:00	10 mg	Daily	Oral
1	Drug Administration-drug A	1	1	1/3/2021 10:00	1/3/2021 10:00	10 mg	Daily	Oral
1	Drug Administration-drug A	1	1	1/4/2021 10:00	1/4/2021 10:00	10 mg	Daily	Oral
1	Drug Administration-drug A	1	1	1/2/2021 10:00	1/2/2021 10:00	10 mg	Daily	Oral
1	Drug Administration-drug A	1	1	1/3/2021 10:00	1/3/2021 10:00	10 mg	Daily	Oral
1	Drug Administration-drug A	1	1	1/4/2021 10:00	1/4/2021 10:00	10 mg	Daily	Oral
1	Drug Administration-drug A	1	1	1/2/2021 10:00	1/2/2021 10:00	10 mg	Daily	Oral
1	Drug Administration-drug A	1	1	1/3/2021 10:00	1/3/2021 10:00	10 mg	Daily	Oral
1	Drug Administration-drug A	1	1	1/4/2021 10:00	1/4/2021 10:00	10 mg	Daily	Oral

RAW.CM

	H	I	J	K	L	M	N	O	P	Q	R	S	T	U	V	W	X	Y
1	FORMID	FSEQ	IGSEQ	CMYN	CMTRT	CMSTDAT	CMENDAT	CMONGO	CMROUTE	CMDECOD	CMDOSFRM	CMDOSFRQ	CMDOSU	DOSUO	DOSFRMO	DOSFRQO	ROUTEO	CMINDC
2	CM	1	1	Y	PRUNE JUICE	1/6/2021	1/6/2021		Oral	PRUNELLA	Liquid	Daily	Liter per Minute					CONSTIPATION
3	CM	1	1	Y	PRUNE JUICE	1/6/2021	1/6/2021		Oral	PRUNELLA	Liquid	Daily	Liter per Minute					CONSTIPATION
4	CM	1	1	N														
5	CM	1	1	N														
6	CM	1	1	N														
7	CM	1	1	Y	PRUNE JUICE	1/7/2021	1/7/2021		Oral	PRUNELLA	Liquid	Daily	Liter per Minute					CONSTIPATION
8	CM	1	1	N														
9	CM	1	1	N														
10	CM	1	1	Y	PRUNE JUICE	1/9/2021	1/9/2021		Oral	PRUNELLA	Liquid	Daily	Liter per Minute					CONSTIPATION
11	CM	1	1	N														

RAM.EG

	A	B	C	D	E	F	G	H	I	J	K	L	M
1	STUDYID	COUNTRY	SITEID	SUBJID	EVENT	EVENTDT	ESEQ	FORM	FORMEID	FSEQ	IGSEQ	EGDAT	EGMETHOD
2	ABC-001	United States	001	001	TREATMENT 1 DAY 1		1	Electrocardiogram	1	1		1/2/2021 9:30	12 Lead Standard
3	ABC-001	United States	001	001	TREATMENT 1 DAY 2		1	Electrocardiogram	1	1		1/3/2021 9:30	12 Lead Standard
4	ABC-001	United States	001	001	TREATMENT 1 DAY 3		1	Electrocardiogram	1	1		1/4/2021 9:30	12 Lead Standard
5	ABC-001	United States	001	002	TREATMENT 1 DAY 1		1	Electrocardiogram	1	1		1/2/2021 9:30	12 Lead Standard
6	ABC-001	United States	001	002	TREATMENT 1 DAY 2		1	Electrocardiogram	1	1		1/3/2021 9:30	12 Lead Standard
7	ABC-001	United States	001	002	TREATMENT 1 DAY 3		1	Electrocardiogram	1	1		1/4/2021 9:30	12 Lead Standard
8	ABC-001	United States	001	003	TREATMENT 1 DAY 1		1	Electrocardiogram	1	1		1/2/2021 9:30	12 Lead Standard
9	ABC-001	United States	001	003	TREATMENT 1 DAY 2		1	Electrocardiogram	1	1		1/3/2021 9:30	12 Lead Standard
10	ABC-001	United States	001	003	TREATMENT 1 DAY 3		1	Electrocardiogram	1	1		1/4/2021 9:30	12 Lead Standard
11	ABC-001	United States	001	004	TREATMENT 1 DAY 1		1	Electrocardiogram	1	1		1/2/2021 9:30	12 Lead Standard
12	ABC-001	United States	001	004	TREATMENT 1 DAY 2		1	Electrocardiogram	1	1		1/3/2021 9:30	12 Lead Standard
13	ABC-001	United States	001	004	TREATMENT 1 DAY 3		1	Electrocardiogram	1	1		1/4/2021 9:30	12 Lead Standard
14	ABC-001	United States	001	005	TREATMENT 1 DAY 1		1	Electrocardiogram	1	1		1/2/2021 9:30	12 Lead Standard
15	ABC-001	United States	001	005	TREATMENT 1 DAY 2		1	Electrocardiogram	1	1		1/3/2021 9:30	12 Lead Standard

16	ABC-001	United States	001	005	TREATMENT 1 DAY 3	1 Electrocardiogram	1	1	1/4/2021 9:30	12 Lead Standard
17	ABC-001	United States	001	006	TREATMENT 1 DAY 1	1 Electrocardiogram	1	1	1/2/2021 9:30	12 Lead Standard
18	ABC-001	United States	001	006	TREATMENT 1 DAY 2	1 Electrocardiogram	1	1	1/3/2021 9:30	12 Lead Standard
19	ABC-001	United States	001	006	TREATMENT 1 DAY 3	1 Electrocardiogram	1	1	1/4/2021 9:30	12 Lead Standard
20	ABC-001	United States	001	007	TREATMENT 1 DAY 1	1 Electrocardiogram	1	1	1/2/2021 9:30	12 Lead Standard
21	ABC-001	United States	001	007	TREATMENT 1 DAY 2	1 Electrocardiogram	1	1	1/3/2021 9:30	12 Lead Standard
22	ABC-001	United States	001	007	TREATMENT 1 DAY 3	1 Electrocardiogram	1	1	1/4/2021 9:30	12 Lead Standard
23	ABC-001	United States	001	008	TREATMENT 1 DAY 1	1 Electrocardiogram	1	1	1/2/2021 9:30	12 Lead Standard
24	ABC-001	United States	001	008	TREATMENT 1 DAY 2	1 Electrocardiogram	1	1	1/3/2021 9:30	12 Lead Standard
25	ABC-001	United States	001	008	TREATMENT 1 DAY 3	1 Electrocardiogram	1	1	1/4/2021 9:30	12 Lead Standard
26	ABC-001	United States	001	009	TREATMENT 1 DAY 1	1 Electrocardiogram	1	1	1/2/2021 9:30	12 Lead Standard
27	ABC-001	United States	001	009	TREATMENT 1 DAY 2	1 Electrocardiogram	1	1	1/3/2021 9:30	12 Lead Standard
28	ABC-001	United States	001	009	TREATMENT 1 DAY 3	1 Electrocardiogram	1	1	1/4/2021 9:30	12 Lead Standard
29	ABC-001	United States	001	010	TREATMENT 1 DAY 1	1 Electrocardiogram	1	1	1/2/2021 9:30	12 Lead Standard
30	ABC-001	United States	001	010	TREATMENT 1 DAY 2	1 Electrocardiogram	1	1	1/3/2021 9:30	12 Lead Standard
31	ABC-001	United States	001	010	TREATMENT 1 DAY 3	1 Electrocardiogram	1	1	1/4/2021 9:30	12 Lead Standard

A	N	O	P	Q	R	S	T	U	V	W	X	Y	Z
1	EGHRMN_EGHRMN_EGORRESU		PRSB_EGORRES PRSB_EGORRESU		QRSAG_EGORRES QRSAG_EGORRESU		QTAG_EGORRES QTAG_EGORRESU		QTCAAG_EGORRES	QTCAAG_EGORRESU	INTP_EGORRES	EGORRES EGCLSIG	
2	50 beats/min		150 msec		88 msec		409 msec		389 msec		NORMAL		
3	56 beats/min		140 msec		100 msec		387 msec		365 msec		NORMAL		
4	55 beats/min		150 msec		91 msec		399 msec		376 msec		NORMAL		
5	53 beats/min		152 msec		86 msec		410 msec		367 msec		NORMAL		
6	56 beats/min		140 msec		99 msec		390 msec		358 msec		NORMAL		
7	55 beats/min		152 msec		91 msec		399 msec		400 msec		NORMAL		
8	53 beats/min		152 msec		97 msec		410 msec		389 msec		NORMAL		
9	55 beats/min		140 msec		100 msec		390 msec		380 msec		NORMAL		
10	55 beats/min		150 msec		91 msec		399 msec		367 msec		NORMAL		
11	53 beats/min		152 msec		97 msec		410 msec		358 msec		NORMAL		
12	55 beats/min		144 msec		98 msec		386 msec		400 msec		NORMAL		
13	55 beats/min		150 msec		91 msec		398 msec		389 msec		NORMAL		
14	55 beats/min		152 msec		97 msec		412 msec		380 msec		NORMAL		
15	55 beats/min		146 msec		98 msec		386 msec		400 msec		NORMAL		

16	55 beats/min	150 msec	91 msec	398 msec	389 mesc	NORMAL
17	67 beats/min	152 msec	97 msec	390 msec	380 msec	NORMAL
18	68 beats/min	145 msec	98 msec	386 msec	367 mesc	NORMAL
19	65 beats/min	153 msec	91 msec	396 msec	358 msec	NORMAL
20	67 beats/min	154 msec	97 msec	398 msec	400 mesc	NORMAL
21	67 beats/min	146 msec	98 msec	390 msec	389 mesc	NORMAL
22	66 beats/min	154 msec	91 msec	396 msec	380 msec	NORMAL
23	67 beats/min	152 msec	97 msec	398 msec	358 mesc	NORMAL
24	67 beats/min	144 msec	90 msec	398 msec	400 mesc	NORMAL
25	56 beats/min	155 msec	91 msec	396 msec	389 mesc	NORMAL
26	68 beats/min	152 msec	98 msec	398 msec	365 mesc	NORMAL
27	66 beats/min	142 msec	90 msec	398 msec	376 mesc	NORMAL
28	58 beats/min	154 msec	90 msec	397 msec	367 mesc	NORMAL
29	68 beats/min	152 msec	98 msec	386 msec	358 mesc	NORMAL
30	66 beats/min	148 msec	92 msec	398 msec	400 mesc	NORMAL
31	58 beats/min	154 msec	90 msec	397 msec	389 mesc	NORMAL

RAM.LB

	A	B	C	D	E	F	G	H	I	J	K	L	M
1	STUDYID	COUNTRY	SITEID	SUBJID	EVENT	LBNAM	LBYN	LBDAT	LBTIM	LBFAST	LBTEST	LBTESTCD	LBCAT
2	ABC-001	United States	001	001	TREATMENT 1 DAY 1	LB Diagnostics - AA	Y	2-Jan-21	9:45	Y	Alkaline Phosphatase	ALP	CHEMISTRY
3	ABC-001	United States	001	001	TREATMENT 1 DAY 1	LB Diagnostics - AA	Y	2-Jan-21	9:45	Y	Calcium	CA	CHEMISTRY
4	ABC-001	United States	001	002	TREATMENT 1 DAY 1	LB Diagnostics - AA	Y	2-Jan-21	9:45	Y	Alkaline Phosphatase	ALP	CHEMISTRY
5	ABC-001	United States	001	002	TREATMENT 1 DAY 1	LB Diagnostics - AA	Y	2-Jan-21	9:45	Y	Calcium	CA	CHEMISTRY
6	ABC-001	United States	001	003	TREATMENT 1 DAY 1	LB Diagnostics - AA	Y	2-Jan-21	9:45	Y	Alkaline Phosphatase	ALP	CHEMISTRY
7	ABC-001	United States	001	003	TREATMENT 1 DAY 1	LB Diagnostics - AA	Y	2-Jan-21	9:45	Y	Calcium	CA	CHEMISTRY
8	ABC-001	United States	001	004	TREATMENT 1 DAY 1	LB Diagnostics - AA	Y	2-Jan-21	9:45	Y	Alkaline Phosphatase	ALP	CHEMISTRY
9	ABC-001	United States	001	004	TREATMENT 1 DAY 1	LB Diagnostics - AA	Y	2-Jan-21	9:45	Y	Calcium	CA	CHEMISTRY
10	ABC-001	United States	001	005	TREATMENT 1 DAY 1	LB Diagnostics - AA	Y	2-Jan-21	9:45	Y	Alkaline Phosphatase	ALP	CHEMISTRY
11	ABC-001	United States	001	005	TREATMENT 1 DAY 1	LB Diagnostics - AA	Y	2-Jan-21	9:45	Y	Calcium	CA	CHEMISTRY
12	ABC-001	United States	001	006	TREATMENT 1 DAY 1	LB Diagnostics - AA	Y	2-Jan-21	9:45	Y	Alkaline Phosphatase	ALP	CHEMISTRY
13	ABC-001	United States	001	006	TREATMENT 1 DAY 1	LB Diagnostics - AA	Y	2-Jan-21	9:45	Y	Calcium	CA	CHEMISTRY
14	ABC-001	United States	001	007	TREATMENT 1 DAY 1	LB Diagnostics - AA	Y	2-Jan-21	9:45	Y	Alkaline Phosphatase	ALP	CHEMISTRY
15	ABC-001	United States	001	007	TREATMENT 1 DAY 1	LB Diagnostics - AA	Y	2-Jan-21	9:45	Y	Calcium	CA	CHEMISTRY
16	ABC-001	United States	001	008	TREATMENT 1 DAY 1	LB Diagnostics - AA	Y	2-Jan-21	9:45	Y	Alkaline Phosphatase	ALP	CHEMISTRY
17	ABC-001	United States	001	008	TREATMENT 1 DAY 1	LB Diagnostics - AA	Y	2-Jan-21	9:45	Y	Calcium	CA	CHEMISTRY
18	ABC-001	United States	001	009	TREATMENT 1 DAY 1	LB Diagnostics - AA	Y	2-Jan-21	9:45	Y	Alkaline Phosphatase	ALP	CHEMISTRY
19	ABC-001	United States	001	009	TREATMENT 1 DAY 1	LB Diagnostics - AA	Y	2-Jan-21	9:45	Y	Calcium	CA	CHEMISTRY
20	ABC-001	United States	001	010	TREATMENT 1 DAY 1	LB Diagnostics - AA	Y	2-Jan-21	9:45	Y	Alkaline Phosphatase	ALP	CHEMISTRY

N	O	P	Q	R	S	T	U	V	W	X
LBORRES	LBORRESU	LBNRIND	LBORNRLO	LBORNRHI	LBSTRESC	LBSTRESN	LBSTRESU	LBSTNRLO	LBSTNRHI	LBCLSIG
62	U/L	NORMAL	50	100	62	62	U/L	50	100	
9	mg/dl	NORMAL	8.5	10	9	9	mg/dl	8.5	10	
65	U/L	NORMAL	50	100	65	65	U/L	50	100	
9.5	mg/dl	NORMAL	8.5	10	9.5	9.5	mg/dl	8.5	10	
66	U/L	NORMAL	50	100	66	66	U/L	50	100	
9.2	mg/dl	NORMAL	8.5	10	9.2	9.2	mg/dl	8.5	10	
67	U/L	NORMAL	50	100	67	67	U/L	50	100	
9.4	mg/dl	NORMAL	8.5	10	9.4	9.4	mg/dl	8.5	10	
64	U/L	NORMAL	50	100	64	64	U/L	50	100	
9.5	mg/dl	NORMAL	8.5	10	9.5	9.5	mg/dl	8.5	10	
63	U/L	NORMAL	50	100	63	63	U/L	50	100	
9.2	mg/dl	NORMAL	8.5	10	9.2	9.2	mg/dl	8.5	10	
66	U/L	NORMAL	50	100	66	66	U/L	50	100	
9.3	mg/dl	NORMAL	8.5	10	9.3	9.3	mg/dl	8.5	10	
60	U/L	NORMAL	50	100	60	60	U/L	50	100	
9	mg/dl	NORMAL	8.5	10	9	9	mg/dl	8.5	10	
62	U/L	NORMAL	50	100	62	62	U/L	50	100	
9.2	mg/dl	NORMAL	8.5	10	9.2	9.2	mg/dl	8.5	10	
63	U/L	NORMAL	50	100	63	63	U/L	50	100	

A.2 SDTM Programming Section
SDTM.DM

```
/*Begin writing SAS program dm.sas*/

/*show structure of the raw DM data set*/

libname RAW ".../directory/sdtm_raw"

/*PROC IMPORT to import raw Demographics(DM) data in excel to SAS*/
PROC IMPORT OUT= RAW.DM DATAFILE= ".../directory/sdtm_raw.xlsx"
    DBMS=xlsx REPLACE;
    SHEET="DM";
  GETNAMES=YES;
RUN;

/*Begin writing SAS program dm.sas*/

/*show structure of the raw DM data set*/
proc contents data=RAW.DM;
run;

/*Create the 1st set of DM variables using existing variables from RAW.DM*/
data DM1;

/*Specify length for standard variables*/
  length STUDYID ARMCD $20  ETHNIC $60 SEX $2 COUNTRY $4 BRTHDTC $20 RACE $100 ;
  set RAW.DM (rename=(COUNTRY=COUNTRY_ SEX=SEX_ AGEU=AGEU_ ETHNIC=ETHNIC_));

/*Derive SITEID, BRTHDTC and COUNTRY*/

  SITEID=SITE;
  BRTHDTC=put(BRTHDAT,yymmdd10.);
  if COUNTRY_="United States" then COUNTRY="USA";
 /*Derive SEX*/
  if SEX_="Female" then SEX="F";
   else if SEX_="Male" then SEX="M";
   else if SEX_="Unknown" then SEX="U";
   else if SEX_="Undifferentiated" then SEX="UNDIFFERENTIATED";

/*Derive ETHNIC AND AGEU*/

  ETHNIC=upcase(ETHNIC_);
  AGEU=upcase(AGEU_);


/*Derive RACE*/

  if cmiss(RACE_WHITE, RACE_BLACK, RACE_HAWAIIAN, RACE_ASIAN, RACE_
AINDIAN,RACE_NOREPORT, RACE_UNKNOWN, RACE_OTHER)=7 then do;
    if not missing(RACE_AINDIAN) then RACE="AMERICAN INDIAN OR ALASKA
AMERICAN";
    else  if not missing(RACE_ASIAN) then RACE="ASIAN";
    else  if not missing(RACE_BLACK) then RACE="BLACK OR AFRICAN AMERICAN";
    else  if not missing(RACE_HAWAIIAN) then RACE="NATIVE HAWAIIAN OR OTHER
PACIFIC ISLANDERS";
```

```
      else  if not missing(RACE_WHITE) then RACE="WHITE";
      else  if not missing(RACE_NOREPORT) then RACE="NOT REPORTED";
      else  if not missing(RACE_UNKNOWN) then RACE="UNKNOWN";
      else  if not missing(RACE_OTHER) then RACE=" ";
   end;

   else if 8-cmiss(RACE_WHITE, RACE_BLACK, RACE_HAWAIIAN, RACE_ASIAN,  RACE_
AINDIAN,RACE_NOREPORT, RACE_UNKNOWN, RACE_OTHER)>1 then RACE="MULTIPLE";
   /*Create SUPPDM Domain*/

   if RACE="MULTIPLE" then do;
     if not missing(RACE_AINDIAN) then RACE1="AMERICAN INDIAN OR ALASKA
AMERICAN";
     else if not missing(RACE_ASIAN) then RACE2="ASIAN";
     else if not missing(RACE_BLACK) then RACE3="BLACK OR AFRICAN AMERICAN";
     else if not missing(RACE_HAWAIIAN) then RACE4="NATIVE HAWAIIAN OR OTHER
PACIFIC ISLANDERS";
     else if not missing(RACE_WHITE) then RACE5="WHITE";
     else if not missing(RACE_NOREPORT) then RACE6="NOT REPORTED";
     else if not missing(RACE_UNKNOWN) then RACE7="UNKNOWN";
     else if not missing(RACE_OTHER) then RACE8=" ";
   end;

   if not missing(race_other) then RACEOTH="OTHER";
     ARMCD="DRUG A";
run;

/*Dropping records with the same ARMCD in order to merge back with DM
variables}*/
/*import TA domain in order to catch variable ARM from TA*/
PROC IMPORT OUT= RAW.TA DATAFILE= ".../directory/sdtm_raw.xlsx"
    DBMS=xlsx REPLACE;
    SHEET="TA";
    GETNAMES=YES;
RUN;

/*Remove duplicate records with the same ARM)*/
proc sort data=RAW.TA out=TA(keep=armcd arm) nodupkey;
      by ARMCD;
run;

/*Merge DM1 with TA domain using ARMCD*/
proc sql;
      create table DM2 as select a.*,b.ARM length 200 from DM1 a left join
TA b on a.ARMCD=b.ARMCD;
quit;

PROC IMPORT OUT=RAW.EX DATAFILE= ".../directory/sdtm_raw.xlsx"
    DBMS=xlsx REPLACE;
    SHEET="EX";
    GETNAMES=YES;
RUN;
```

```
/*Create RFSTDTC and RFENDTC from EX domain*/

data EX1;
   set RAW.EX(keep=SUBJID EXSTDAT);
   EXDTS=datepart(EXSTDAT);
   EXTMS=timepart(EXSTDAT);
   EXDTS_DT=put(EXDTS,yymmdd10.);
   EXDTS_TM=put(EXTMS,time8.);
   EXSTDTC=strip(EXDTS_DT)||"T"||strip(EXDTS_TM);
run;

data EX2(rename=(EXSTDTC=RFSTDTC))
     EX3(rename=(EXSTDTC=RFENDTC));
  set EX1;
  by SUBJID EXSTDTC;
  if first.SUBJID then output EX2;
  if last.SUBJID then output EX3;
run;

proc sql;
    create table DM3 as select a.*,b.RFSTDTC from DM2 a left join EX2 b on
a.SUBJID=b.SUBJID;
  create table DM4 as select a.*,b.RFENDTC from DM3 a left join EX3 b on
a.SUBJID=b.SUBJID;
quit;


data Final;
/*Defining DOMAIN, STUDYID, USUBJID, ACTARM, ACTARMCD*/
set DM4;
  length ACTARMCD ARMCD $20. ARM ACTARM $200.;
  DOMAIN="DM";
  STUDYID="ABC-001";
  USUBJID=STRIP(STUDYID)||"-"||STRIP(SITEID)||"-"||STRIP(SUBJID);
  ACTARM=strip(ARM);
  ACTARMCD = ARMCD;
  format _all_;
  informat _all_;
run;


libname SDTM ".../directory";
data SDTM.DM(label="Demographics");

/*Assign variable attributes such as label and length to conform with
SDTM.DM Specification (these will also be the same attributes as the SDTM
IG).*/
attrib
       STUDYID        label = "Study Identifier"                 length = $20
       DOMAIN         label = "Domain Abbreviation"              length = $2
       USUBJID        label = "Unique Subject Identifier"        length = $40
       SUBJID         label = "Subject Identifier for the Study" length = $20
       RFSTDTC        label = "Subject Reference Start Date/Time" length = $20
       RFENDTC        label = "Subject Reference End Date/Time"  length = $20
       BRTHDTC        label = "Date/Time of Birth"               length = $20
```

```
          SITEID          label = "Study Site Identifier"              length = $10
          AGE             label = "Age"                                length = 8
          AGEU            label = "Age Units"                          length = $10
          SEX             label = "Sex"                                length = $2
          RACE            label = "Race"                               length = $100
          ETHNIC          label = "Ethnicity"                          length = $60
          ARM             label = "Description of Planned Arm"         length = $200
          ARMCD           label = "Planned Arm Code"                   length = $20
          ACTARMCD        label = "Actual Arm Code"                    length = $20
          ACTARM          label = "Description of Actual Arm"          length = $200
          COUNTRY         label = "Country"                            length = $4
          ;
     set Final;
     keep STUDYID DOMAIN USUBJID SUBJID RFSTDTC RFENDTC BRTHDTC SITEID AGE AGEU
SEX RACE ETHNIC ARMCD ARM ACTARMCD ACTARM COUNTRY
          ;
run;
```

SDTM.SUPPDM

```
/*suppdm*/

data SUPPDM;
/*Create SUPPxx Variables which will always be QNAM, QLABEL, QVAL, QORIG,
IDVAR, IDVARVAL and RDOMAIN. */
  set Final;
  length RDOMAIN $2. IDVAR $8. QNAM USUBJID IDVARVAL QLABEL $40. QORIG QVAL
$100.;
  RDOMAIN="DM";
  IDVAR="";
  IDVARVAL="";
  QORIG="CRF";
  if RACE="MULTIPLE" and ^missing(RACE_AINDIAN) then do;
    QNAM="RACE1";
    QLABEL="American Indian/Alaska Native";
    QVAL="AMERICAN INDIAN OR ALASKA NATIVE";
  output;
 end;
 if RACE="MULTIPLE"  and ^missing(RACE_ASIAN) then do;
   QNAM="RACE2";
   QLABEL="ASIAN";
   QVAL="ASIAN";
  output;
end;
 if RACE="MULTIPLE" and ^missing(RACE_BLACK) then do;
   QNAM="RACE3";
   QLABEL="Black or African American";
   QVAL="BLACK OR AFRICAN AMERICAN";
  output;
 end;
```

```
if RACE="MULTIPLE" and ^missing(RACE_HAWAIIAN) then do;
  QNAM="RACE4";
  QLABEL="Native Hawaiian/Pacific Islander";
  QVAL="NATIVE HAWAIIAN OR OTHER PACIFIC ISLANDER";
 output;
end;
if RACE="MULTIPLE" and ^missing(RACE_WHITE) then do;
  QNAM="RACE5";
  QLABEL="White";
  QVAL="WHITE";
  output;
end;
if RACE="MULTIPLE" and ^missing(RACE_NOREPORT) then do;
  QNAM="RACE6";
  QLABEL="Not reported";
  QVAL="NOT REPORTED";
 output;
end;
if RACE="MULTIPLE" and ^missing(RACE_UNKNOWN) then do;
  QNAM="RACE7";
  QLABEL="Unknown";
  QVAL="UNKNOWN";
  output;
end;
if RACE="MULTIPLE" and ^missing(RACE_OTHER) then do;
  QNAM="RACE8";
  QLABEL="Other";
  QVAL="OTHER";
  output;
end;
 QORIG="CRF";

run;

libname SDTM ".../directory";
data SDTM.SUPPDM(label="Supplemental Qualifiers for DM");

/*Assign variable attributes such as label and length to conform with
SDTM.SUPPDM Specification (these will also be the same attributes as the
SDTM IG).*/

attrib
      STUDYID       label = "Study Identifier"                 length = $20
      RDOMAIN       label = "Related Domain Abbreviation"       length = $2
      USUBJID       label = "Unique Subject Identifier"         length = $40
      IDVAR         label = "Identifying Variable"              length = $8
      IDVARVAL      label = "Identifying Variable Value"        length = $40
      QNAM          label = "Qualifier Variable Name"           length = $40
      QLABEL        label = "Qualifier Variable Label"          length = $40
      QVAL          label = "Data Value"                        length = $100
      QORIG         label = "Origin"                            length = $100
      ;
```

```
  set SUPPDM;
   keep STUDYID RDOMAIN USUBJID IDVAR IDVARVAL QNAM QLABEL QVAL QORIG
      ;
run;
```

SDTM.DS

```
/*create a libref called RAW pointing to the pathway under E drive*/
libname RAW ".../directory/sdtm_raw";
/*PROC IMPORT to import raw ENDDOSE data in excel to SAS*/
PROC IMPORT OUT= RAW.DS DATAFILE= ".../directory/sdtm_raw.xlsx"
     DBMS=xlsx REPLACE;
     SHEET="ENDDOSE";
     GETNAMES=YES;
RUN;

/*Begin writing SAS program ds.sas*/
/*show structure of the raw  DS data set*/
proc contents data=RAW.DS;
run;

/*termination- end of dosing*/

/*Create the 1st set of DS variables using existing variables from RAW.
DS*/

data DS1;
   set RAW.DS;
   /*Define DOMAIN, STUDYID, SITEID,USUBJID,DSSTDTC*/
   DOMAIN="DS";
   STUDYID="ABC-001";
   SITEID=SITE;
   USUBJID=STRIP(STUDYID)||"-"||STRIP(SITEID)||"-"||STRIP(SUBJID);
   DSCAT="DISPOSITION EVENT";
   DSSTDTC=strip(put(DSSTDAT,yymmdd10.));
   /*Derive DSTERM*/
   if upcase(DSDECOD)="COMPLETED" then DSTERM="COMPLETED";
   else if upcase(DSDECOD)="ADVERSE EVENT" then DSTERM="ADVERSE EVENT";
   else if upcase(DSDECOD)="DEATH" then DSTERM="DEATH";
   else if upcase(DSDECOD)="Lost To Follow-Up" then DSTERM="Lost To
Follow-Up";
   else if upcase(DSDECOD)="PREGANCY" then DSTERM="PREGANCY";
   else if upcase(DSDECOD)="PROGRESSIVE DISEASE" then DSTERM="PROGRESSIVE
DISEASE";
   else if upcase(DSDECOD)="PROTOCOL DEVIATION" then DSTERM="PROTOCOL
DEVIATION";
   else if upcase(DSDECOD)="SCREEN FAILURE" then DSTERM="SCREEN FAILURE";
   else if upcase(DSDECOD)="SITE TERMINATED BY SPONSOR" then DSTERM="SITE
TERMINATED BY SPONSOR";
   else if upcase(DSDECOD)="STUDY TERMINATED BY SPONSOR" then DSTERM="STUDY
TERMINATED BY SPONSOR";
```

```
     else if upcase(DSDECOD)="WITHDRAWN BY SUBJECT" then DSTERM="WITHDRAWN BY
SUBJECT";
     else if upcase(DSDECOD)="OTHER" then DSTERM="OTHER";
run;
/*Sort data set DS1 by USUBJID, DSSTDTC, DSDECOD*/
proc sort data=DS1 out=DS2;
  by USUBJID DSSTDTC DSDECOD;
run;
/*Derive DSSEQ*/
data Final;
  set DS2;
  length DSSEQ 8.;
  by USUBJID DSSTDTC DSDECOD;
  if FIRST.USUBJID then DSSEQ=0;
  DSSEQ+1;
  output;
  format _all_;
  informat _all_;
run;
libname SDTM ".../directory";
data SDTM.DS (label="Disposition");
/*Assign variable attributes such as label and length to conform with
SDTM.DS Specification (these will also be the same attributes as the SDTM
IG).*/
attrib
 STUDYID      label = "Study Identifier"                      length = $20
 DOMAIN       label = "Domain Abbreviation"                   length = $2
 USUBJID      label = "Unique Subject Identifier"             length = $40
 DSSEQ        label = "Sequence Number"                       length = 8
 DSTERM       label = "Reported Term for the Disposition Event" length = $200
 DSDECOD      label = "Standardized Disposition Term"         length = $200
 DSCAT        label = "Category for Disposition Event"        length = $40
 DSSTDTC      label = "Start Date/Time of Disposition Event"  length = $20
   ;
  set Final;
   keep STUDYID DOMAIN USUBJID DSSEQ DSTERM DSDECOD DSCAT DSSTDTC
   ;
```

SDTM.AE

```
/*create a libref called RAW pointing to the pathway under E drive*/
libname RAW ".../directory/sdtm_raw";

/*PROC IMPORT to import raw Adverse Events(AE) data in excel to SAS*/
PROC IMPORT OUT= RAW.AE DATAFILE= ".../directory/sdtm_raw.xlsx"
     DBMS=xlsx REPLACE;
     SHEET="AE";
     GETNAMES=YES;
```

```
RUN;

/*Begin writing SAS program ae.sas*/

/*show structure of the raw AE dataset*/
proc contents data=RAW.AE;
run;

/*Assign the character value from "Mild", "Moderate" and "Severe" to char-
acter value "1", "2" and "3" for variable AETOXGR with PROC FORMAT*/
proc format;
 value $AETOXGR
 "Mild"="1"
 "Moderate"="2"
 "Severe"="3"
;

quit;

data AE1;

/*Specify length for standard variables*/
length STUDYID AESTDTC AEENDTC AEBODSYS $20
       DOMAIN AESER AESCONG AESDISAB AESDTH AESHOSP AESMIE AETOXGR $2
       USUBJID AEENRTPT AEENTPT $40
       AELLTCD AEPTCD AEHLTCD AEHLGTCD 8
       AETERM AEDECOD AEHLT AEHLGT $200
       AELLT $100
       AEACN AEREL AEOUT $50;
  set RAW.AE(rename=(AETERM=AETERM_ AEACN=AEACN_ AESER=AESER_ AEREL=AEREL_
AEOUT=AEOUT_ AESCONG=AESCONG_ AESDISAB=AESDISAB_ AESDTH=AESDTH_
AESHOSP=AESHOSP_ AESMIE=AESMIE_));
  DOMAIN="AE";
  STUDYID="ABC-001";
  USUBJID=STRIP(STUDYID)||"-"||STRIP(SITEID)||"-"||STRIP(SUBJID);

   /*Derive AETERM, AELLT, AELLTCD, AEDECOD, AEPTCD, AEHLT, AEHLTCD, AEHL-
GT, AEHLGTCD, AEBODSYS, AEACN, AEOUT*/
  AETERM=strip(upcase(AETERM_));
  AELLT=strip(upcase(LLT));
  AELLTCD=LLTCD;
  AEDECOD=strip(upcase(PT));
  AEPTCD=PT_CD;
  AEHLT=strip(upcase(HLT));
  AEHLTCD=HLTCD;
  AEHLGT=strip(upcase(HLGT));
  AEHLGTCD=HLGTCD;
  AEBODSYS=strip(upcase(SOC));
  AEACN=strip(upcase(AEACN_));
  AEOUT=strip(upcase(AEOUT_));

/*Derive AESER, AESCONG, AESDISAB, AESDTH, AESHOSP, AESMIE */
  if AESER_="Yes" then AESER="Y";
  else if AESER_="No" then AESER="N";
```

```
  if AEREL_="Yes" then AEREL="Y";
  else if AEREL_="No" then AEREL="N";
  if AESCONG_="Yes" then AESCONG="Y";
  else if AESCONG_="No" then AESCONG="N";
  if AESDISAB_="Yes" then AESDISAB="Y";
  else if AESDISAB_="No" then AESDISAB="N";
  if AESDTH_="Yes" then AESDTH="Y";
  else if AESDTH_="No" then AESDTH="N";
  if AESHOSP_="Yes" then AESHOSP="Y";
  else if AESHOSP_="No" then AESHOSP="N";
  if AESMIE_="Yes" then AESMIE="Y";
  else if AESMIE_="No" then AESMIE="N";
  /*Format AETOXGR, AESTDTC, AEENDTC */
  AETOXGR=put(AESEV, AETOXGR.);
  AESTDTC=put(AESDAT,yymmdd10.);
  AEENDTC=put(AEENDAT,yymmdd10.);
  /*Derive AEENRTPT*/
  if missing(AEENDTC) and AEOUT = "NOT RECOVERED OR NOT RESOLVED" then
AEENRTPT = "ONGOING";
  else if missing(AEENDTC) and AEOUT = "UNKNOWN" then AEENRTPT = "UNKNOWN";
  if AEENRTPT in("ONGOING", "UNKNOWN") then AEENTPT = "END OF STUDY";
  keep STUDYID AESTDTC AEENDTC DOMAIN AESER AESCONG AESDISAB AESDTH AESHOSP
AEBODSYS
     USUBJID AEENRTPT AEENTPT AELLTCD AEPTCD AEHLTCD AEHLGTCD AETERM AEDECOD
AEHLT AEHLGT
     AELLT AEACN AEREL AEOUT AETOXGR AESMIE;
run;

/*Sort data set AE1 by USUBJID, AESTDTC, AEENDTC, AETERM */
proc sort data=AE1 out=AE2;
  by USUBJID AESTDTC AEENDTC AETERM;
run;

/*Derive AESEQ*/
data FINAL;
  set AE2;
  by USUBJID AESTDTC AEENDTC AETERM;
  if FIRST.USUBJID then AESEQ=0;
    AESEQ+1;
  output;
  format _all_;
  informat _all_;
run;
libname SDTM"../directory";
data SDTM.AE(label="Adverse Events");
/*Assign variable attributes such as label and length to conform with
SDTM.AE Specification (these will also be the same attributes as the SDTM
IG).*/

attrib
 STUDYID       label = "Study Identifier"                 length = $20
 DOMAIN        label = "Domain Abbreviation"              length = $2
 USUBJID       label = "Unique Subject Identifier"        length = $40
```

```
AESEQ          label = "Sequence Number"                          length = 8
AETERM         label = "Reported Term for the Adverse Event"      length = $200
AELLT          label = "Lowest Level Term"                        length = $100
AELLTCD        label = "Lowest Level Term Code"                   length = 8
AEDECOD        label = "Dictionary-Derived Term"                  length = $200
AEPTCD         label = "Preferred Term Code"                      length = 8
AEHLT          label = "High-Level Term"                          length = $200
AEHLTCD        label = "High-Level Term Code"                     length = 8
AEHLGT         label = "High-Level Group Term"                    length = $200
AEHLGTCD       label = "High-Level Group Term Code"               length = 8
AEBODSYS       label = "Body System or Organ Class"               length = $20
AESER          label = "Serious Event"                            length = $2
AEACN          label = "Action Taken with Study Treatment"        length = $50
AEREL          label = "Causality"                                length = $50
AEOUT          label = "Outcome of Adverse Event"                 length = $50
AESCONG        label = "Congenital Anomaly or Birth Defect"       length = $2
AESDISAB       label = "Persist or Significant Disability"        length = $2
AESDTH         label = "Results in Death"                         length = $2
AESHOSP        label = "Requires or Prolongs Hospitalization"     length = $2
AESMIE         label = "Other Medically Important Serious Event"  length = $2
AETOXGR        label = "Standard Toxicity Grade"                  length = $2
AESTDTC        label = "Start Date/Time of Adverse Event"         length = $20
AEENDTC        label = "End Date/Time of Adverse Event"           length = $20
AEENRTPT       label = "End Relative to Reference Time Point"     length = $40
AEENTPT        label = "End Reference Time Point"                 length = $40
    ;
  set FINAL;
    ;
run;
```

SDTM.EX

```
/*create a libref called RAW pointing to the pathway under E drive*/
libname RAW ".../directory/sdtm_raw";

/*PROC IMPORT to import raw Exposure (EX) data in excel to SAS*/
PROC IMPORT OUT= RAW.EX DATAFILE= ".../directory/sdtm_raw.xlsx"
          DBMS=xlsx REPLACE;
     SHEET="EX";
     GETNAMES=YES;
RUN;

/*Begin writing SAS program ex.sas*/
/*show structure of the raw EX data set*/
proc contents data=RAW.EX;
run;

/*Create the 1st set of EX variables using existing variables from RAW.
EX*/
```

```
data EX1;
/*Specify length for standard variables*/
   length exstdtc exendtc $20. EXTRT $100;

   /*Rename EXDOSU, EXDOSE, EXDOSFRQ, EXROUTE, EXREFID */
   set RAW.EX(rename=(EXDOSU=EXDOSU_ EXDOSE=EXDOSE_ EXDOSFRQ=EXDOSFRQ_
EXROUTE=EXROUTE_ EXREFID=EXREFID_));
/*Define DOMAIN, USUBJID, EXCAT, EXDOSFRM */
/*Derive EXTRT, EXDOSU, EXDOSE, EXDOSFRQ, EXROUTE, EXDTS, EXTMS */

   DOMAIN="EX";
   USUBJID=STRIP(STUDYID)||"-"||STRIP(SITEID)||"-"||STRIP(SUBJID);
   if index(upcase(FORM),"DRUG A")>0 then EXTRT="DRUG A";
   EXCAT="DOSING";
   EXDOSU=upcase(strip(EXDOSU_));
   EXDOSE=EXDOSE_;
   EXDOSFRM="TABLET";
   EXDOSFRQ=upcase(strip(EXDOSFRQ_));
   EXROUTE=upcase(strip(EXROUTE_));
   EXDTS=datepart(EXSTDAT);
   EXTMS=timepart(EXSTDAT);
/*Format EXDTS_DT, EXDTS_TM , EXDTE_DT , EXDTE_TM */
/*Derive EXDTE , EXTME , EXSTDTC , EXENDTC */

   EXDTS_DT=put(EXDTS,yymmdd10.);
   EXDTS_TM=put(EXTMS,time8.);

   EXDTE=datepart(EXENDAT);
   EXTME=timepart(EXENDAT);
   EXDTE_DT=put(EXDTE,yymmdd10.);
   EXDTE_TM=put(EXTME,time8.);

   EXSTDTC=strip(EXDTS_DT)||"T"||strip(EXDTS_TM);
   EXENDTC=strip(EXDTE_DT)||"T"||strip(EXDTE_TM);
run;

/*Sort data set EX1 by USUBJID, EXSTDTC, EXENDTC, EXTRT, EXDOSE */
proc sort data=EX1 out=EX2;
   by USUBJID EXSTDTC EXENDTC EXTRT EXDOSE ;
run;

/*Derive EXSEQ*/
data Final;
  set EX2;
  length EXSEQ 8.;
  by USUBJID EXSTDTC EXENDTC EXTRT EXDOSE ;
  if FIRST.USUBJID then EXSEQ=0;
    EXSEQ+1;
  output;
  format _all_;
  informat _all_;
run;
libname SDTM ".../directory";
```

```
data SDTM.EX(label="Exposure");
/*Assign variable attributes such as label and length to conform with
SDTM.EX Specification (these will also be the same attributes as the SDTM
IG).*/
attrib
        STUDYID         label = "Study Identifier"                  length = $20
        DOMAIN          label = "Domain Abbreviation"               length = $2
        USUBJID         label = "Unique Subject Identifier"         length = $40
        EXSEQ           label = "Sequence Number"                   length = 8
        EXREFID         label = "Reference ID"                      length = $20
        EXTRT           label = "Name of Treatment"                 length = $100
        EXCAT           label = "Category of Treatment"             length = $40
        EXDOSE          label = "Dose"                              length = 8
        EXDOSU          label = "Dose Units"                        length = $40
        EXDOSFRM        label = "Dose Form"                         length = $20
        EXDOSFRQ        label = "Dosing Frequency per Interval"     length = $20
        EXROUTE         label = "Route of Administration"           length = $40
        EXSTDTC         label = "Start Date/Time of Treatment"      length = $40
        EXENDTC         label = "End Date/Time of Treatment"        length = $40
        ;
    set Final;
      keep STUDYID DOMAIN USUBJID EXSEQ EXREFID EXTRT EXCAT EXDOSE EXDOSU
EXDOSFRM EXDOSFRQ EXROUTE EXSTDTC EXENDTC;
run;
```

SDTM.CM

```
/*create a libref called RAW pointing to the pathway under E drive*/
libname RAW """.../directory/sdtm_raw";

/*PROC IMPORT to import raw Concomitant Medication data in excel to SAS*/
PROC IMPORT OUT= RAW.CM DATAFILE= ".../directory/sdtm_raw.xlsx"
     DBMS=xlsx REPLACE;
     SHEET="CM";
     GETNAMES=YES;
RUN;

/* Begin writing SAS program cm.sas*/

/*show structure of the raw CM data set*/
proc contents data=RAW.CM;
run;

/*Create the 1st set of CM variables using existing variables from RAW.
CM*/
data CM1;

/*Specify length for standard variables*/
length STUDYID CMSTDTC CMENDTC CMENRTPT $20. DOMAIN $2. USUBJID CMCAT CMROUTE
$40. CMTRT CMINDC $200.;

/*Rename STUDYID, CMTRT, CMINDC, CMDOSU, CMROUTE , CMDOSFRM , CMDOSFRQ,
CMDECOD */
```

```
  set RAW.CM(where=(CMYN="Y") rename=(STUDYID=STUDYID_ CMTRT=CMTRT_
CMINDC=CMINDC_ CMDOSU=CMDOSU_ CMROUTE=CMROUTE_ CMDOSFRM=CMDOSFRM_
CMDOSFRQ=CMDOSFRQ_ CMDECOD=CMDECOD_));
/*Define DOMAIN, CMCAT*/
DOMAIN="CM";
/*Derive STUDYID, USUBJID, CMROUTE, CMDOSU, CMDOSFRM, CMDOSFRQ */
STUDYID=strip(STUDYID_);
USUBJID=STRIP(STUDYID)||"-"||STRIP(SITEID)||"-"||STRIP(SUBJID);
CMTRT=strip(CMTRT_);
CMCAT="GENERAL";

CMROUTE=strip(upcase(CMROUTE_));
CMDOSU=strip(upcase(CMDOSU_));
CMDOSFRM=strip(upcase(CMDOSFRM_));
CMDOSFRQ=strip(upcase(CMDOSFRQ_));
/*Format CMSTDTC, CMENDTC */
CMSTDTC=put(CMSTDAT,yymmdd10.);
CMENDTC=put(CMENDAT,yymmdd10.);
/*Derive CMTRT, CMINDC, CMDECOD, CMENRTPT */
CMTRT=strip(upcase(CMTRT_));
CMINDC=strip(upcase(CMINDC_));
CMDECOD=strip(upcase(CMDECOD_));
if CMONGO^="" then CMENRTPT="ONGOING";
run;
/*Sort data set CM1 by USUBJID, CMCAT, CMSTDTC, CMENDTC, CMTRT */

proc sort data=CM1 out=CM2;
   by USUBJID CMCAT CMSTDTC CMENDTC CMTRT;
run;

/*Derive CMSEQ*/
data Final;
  set CM2;
length CMSEQ 8.;
by USUBJID CMCAT CMSTDTC CMENDTC CMTRT;
if FIRST.USUBJID then CMSEQ=0;
  CMSEQ+1;
output;
run;
libname SDTM "…/directory";
data SDTM.CM(label="Concomitant and Prior Medications");

/*Assign variable attributes such as label and length to conform with SDTM.
CM Specification (these will also be the same attributes as the SDTM IG).*/
Attrib
  STUDYID             label = "Study Identifier"                    length = $20
  DOMAIN              label = "Domain Abbreviation"                 length = $2
  USUBJID             label = "Unique Subject Identifier"           length = $40
  SUBJID              label = "Subject Identifier for the Study" length = $20
  CMSEQ   label = "Sequence Number"                                 length = 8
  CMTRT   label = "Reported name of drug, Medication or Therapy" length = $200
  CMDECOD label = "Standardized Medication Name"                    length = $100
  CMCAT   label = "Category for Medication"                         length = $40
  CMINDC  label = "Indication"                                      length = $200
```

```
  CMDOSU    label = "Dose Units"                                    length = $40
  CMDOSFRM label = "Dose Form"                                      length = $40
  CMDOSFRQ label = "Dose Frequency"                                 length = $40
  CMROUTE   label = "Route of Administration"                       length = $40
  CMSTDTC   label = "Start Date/Time of Medication"                 length = $20
  CMENDTC   label = "End Date/Time of Medication"                   length = $20
  CMENRTPT label = "End Relative to Reference Time Point"           length = $20
  ;
  set Final;
    keep STUDYID DOMAIN USUBJID SUBJID CMSEQ CMTRT CMDECOD CMCAT CMDOSU
CMDOSFRM CMDOSFRQ CMINDC CMROUTE CMSTDTC CMENDTC CMENRTPT
        ;
run;
```

SDTM.EG

```
/*create a libref called RAW pointing to the pathway under E drive*/
libname RAW ".../directory/sdtm_raw";
/*PROC IMPORT to import raw EG data in excel to SAS*/
PROC IMPORT OUT= RAW.EG DATAFILE= ".../directory/sdtm_raw.xlsx"
        DBMS=xlsx REPLACE;
      SHEET="EG";
      GETNAMES=YES;
RUN;
/*Begin writing SAS program eg.sas*/

/*show structure of the raw EG dataset*/
proc contents data=RAW.EG;
run;

/*Assign the character values from "HR", "INTP", "PR", "QRS", "QT", "QTca"
to character values "Heart Rate", "Interpretation", "PR Interval", "QRS
Interval",
"QT Interval" and "QT Interval" for variable EGTESTCD using PROC FORMAT*/

proc format;
  value $EGTESTCD
     "HR"="Heart Rate"
        "INTP"="Interpretation"
        "PR"="PR Interval"
        "QRS"="QRS Interval"
        "QT"="QT Interval"
        "QTca"="QTca Interval"
        ;
quit;
data EG1;

  /*Specify length for standard variables*/
  length INTP $300. USUBJID $40. HR PR QRS QT QTca $200. VISIT $40.;

  /*Rename STUDYID EGMETHOD*/
  set RAW.EG(rename=(STUDYID=STUDYID_ EGMETHOD=EGMETHOD_));
```

```
/*Define DOMAIN*/
DOMAIN="EG";

/*Derive STUDYID , USUBJID , EGREFID */
STUDYID=strip(STUDYID_);
USUBJID=STRIP(STUDYID)||"-"||STRIP(SITEID)||"-"||STRIP(SUBJID);
EGMETHOD=upcase(strip(EGMETHOD_));
EGREFID=strip(put(FSEQ,best.));

/*Derive HR, PR, QRS, QT, QTca, INTP*/

HR=catx("~",Coalescec(put(EGHRMN_EGORRES,best.),"_null_"),EGHRMN_EGORRESU);
PR=catx("~",Coalescec(put(PRSB_EGORRES,best.),"_null_"),PRSB_EGORRESU);
QRS=catx("~",Coalescec(put(QRSAG_EGORRES,best.),"_null_"),QRSAG_EGORRESU);
QT=catx("~",Coalescec(put(QTAG_EGORRES,best.),"_null_"),QTAG_EGORRESU);
QTca=catx("~",Coalescec(put(QTCAAG_EGORRES,best.),"_null_"),QTCAAG_
EGORRESU);
INTP=catx("~",INTP_EGORRES,EGORRES);

/*Derive VISIT*/
if index(upcase(EVENT),"SCREENING") then VISIT="SCREENING";
else if index(upcase(EVENT),"FOLLOW UP") then VISIT="FOLLOW-UP";
else if index(upcase(EVENT),"DAY") then VISIT="DAY
"||strip(substr(upcase(EVENT),length(EVENT)-1));
else VISIT=strip(upcase(EVENT));
run;

/*Sort data set EG1 by STUDYID, DOMAIN, USUBJID, EGREFID, VISIT, EGDAT */
proc sort data=EG1;
   by STUDYID DOMAIN USUBJID EGMETHOD EGREFID VISIT EGDAT ;
run;

/*Transpose data set EG1 with HR, PR, QRS, QT, QTca, INTP */
proc transpose data=EG1 out=EG2;
   by STUDYID DOMAIN USUBJID EGMETHOD EGREFID VISIT EGDAT;
   var HR PR QRS QT QTca INTP;
run;
data EG3;

/*Specify length for standard variables*/
   length EGDTC EGTEST $20. EGTESTCD $8. EGORRESU $40.  EGORRES EGSTRESC
$100.;
   set EG2;

/*Derive EGTESTCD, EGSTRESC, EGORRES, EGCAT, EGPOS, EGORRESU, EGSTRESC
*/
   EGTESTCD=upcase(strip(_name_));
   EGTEST=put(EGTESTCD,$EGTESTCD.);
   if _name_="INTP" then do;
     EGSTRESC=upcase(strip(scan(col1,1,"~")));
     if index(col1,"~") then EGORRES=upcase(strip(scan(col1,2,"~")));
     else EGORRES=EGSTRESC;
end;
   else if _name_^="INTP" then do;
     if EGTESTCD="HR" then EGCAT="MEASUREMENT";
```

```
      else if EGTESTCD in ("PR","QRS","QT","QTCA") then EGCAT="INTERVAL";
      if EGTESTCD in ("HR","PR","QRS","QT","QTCA") then EGPOS="SUPINE";
      EGORRES=strip(scan(col1,1,"~"));
      EGORRESU=strip(scan(col1,2,"~"));
      EGSTRESC=EGORRES;
  end;

    /*Derive EGDTC*/

  EGD=datepart(EGDAT);
  EGM=timepart(EGDAT);
  EGDTC_DT=put(EGD,yymmdd10.);
  EGDTC_TM=put(EGM,tod8.);
  EGDTC=strip(EGDTC_DT)||"T"||strip(EGDTC_TM);
run;

/*Sort data set EG3 by USUBJID*/
proc sort data=EG3;
  by USUBJID;
run;

/*Sort data set SDTM.EX by USUBJID without duplicate values*/
proc sort data=SDTM.EX out=EX(keep=USUBJID EXSTDTC) nodupkey;
  by USUBJID;
run;
data EG4;

/*Merge data set EG3 and EX*/
  merge EG3(in=a) EX(in=b keep=USUBJID EXSTDTC);
  by USUBJID;

  /*Derive EGDTM and EXSTDTM*/
  if length(EGDTC)>=10 then EGDTM=input(substr(EGDTC,1,16),e8601dt.);
  if length(EXSTDTC)>=10 then EXSTDTM=input(substr(EXSTDTC,1,16),e8601dt.);
run;

/*Select the condition when EGDTM is on or before EXSTDTM and nonmissing
EGSTRESC to
create the baseline flag with a value of "Y" for the last nonmissing result
for each EGTESTCD.*/

/*Sort data set EG4 by USUBJID, EGTESTCD,EGDTC */
proc sort data=EG4 out=BASEFL1;
  by USUBJID EGTESTCD EGDTC;
  where (EGDTM<=EXSTDTM and ^missing(EGSTRESC));
run;
data BASEFL2;
  set BASEFL1;

  /*Derive EGBLFL*/
  by USUBJID EGTESTCD EGDTC ;
  if last.EGTESTCD then EGBLFL="Y";
  keep USUBJID EGTESTCD EGDTC EGBLFL;
run;

/*Sort data set BASEFL2 by USUBJID, EGTESTCD, EGDTC */
```

```
proc sort data=BASEFL2;
  by USUBJID EGTESTCD EGDTC;
run;
/*Sort data set EG4 by USUBJID, EGTESTCD, EGDTC */
proc sort data=EG4;
   by USUBJID EGTESTCD EGDTC;
run;
data EG5;
/*Merge data set EG4 and BASEFL2*/
  merge EG4(in=a) BASEFL2(in=b);
  by USUBJID EGTESTCD EGDTC;
  if a;
run;
proc sort data=EG5;
  by VISIT;
run;
proc sort data=SDTM.TV out=TV(keep=VISIT VISITNUM) nodupkey;
  by VISIT;
run;
data EG6;
/*Merge data set EG5 and TV*/
  merge EG5(in=a) TV(in=b);
  by VISIT;
  if a;
run;
/*Sort data set EG6 by USUBJID, EGCAT, EGTESTCD, EGDTC, VISIT */

proc sort data=EG6;
  by USUBJID EGCAT EGTESTCD EGDTC VISIT;
run;
/*Derive EGSEQ*/

data Final;
  set EG6;
    by USUBJID EGCAT EGTESTCD EGDTC VISIT;
    if first.USUBJID then EGSEQ = 0;
      EGSEQ+1;
    output;
  format _all_;
  informat _all_;
run;
libname SDTM ".../directory";
data SDTM.EG(label="Electrocardiogram Test Results");

/*Assign variable attributes such as label and length to conform with
SDTM.EG Specification (these will also be the same attributes as the SDTM
IG).*/

attrib
    STUDYID    label = "Study Identifier"             length = $20
    DOMAIN     label = "Domain Abbreviation"          length = $2
```

```
      USUBJID    label = "Unique Subject Identifier"              length = $40
      EGSEQ      label = "Sequence Number"                        length = 8
      EGREFID    label = "ECG Reference ID"                       length = $20
      EGTESTCD   label = "ECG Test or Examination Short Name"      length = $20
      EGTEST     label = "ECG Test or Examination Name"            length = $20
      EGCAT      label = "Category for ECG"                        length = $20
      EGPOS      label = "ECG POSITION OF SUBJECTS"                length = $20
      EGORRES    label = "Result or Finding in Original Units"     length = $100
      EGORRESU   label = "Original Units"                          length = $40
      EGSTRESC   label = "Character Result/Finding in Std Format"  length = $100
      EGMETHOD   label = "Method of ECG Test"                      length = $40
      EGBLFL     label = "Baseline Flag"                           length = $60
      VISIT      label = "Visit Name"                              length = $40
      VISITNUM   label = "Visit Number"                            length = 8
      EGDTC      label = "Date/Time of ECG"                        length = $40

   ;
  set Final;
   keep STUDYID DOMAIN USUBJID EGSEQ EGREFID EGTESTCD EGTEST EGCAT EGPOS
EGORRES EGORRESU EGSTRESC EGMETHOD EGBLFL VISIT VISITNUM EGDTC
      ;
run;
```

SDTM.LB

```
/*create a libref called RAW pointing to the pathway under E drive*/
libname RAW ".../directory/sdtm_raw";

/*PROC IMPORT to import raw EG data in excel to SAS*/
PROC IMPORT OUT= RAW.LB DATAFILE= ".../directory/sdtm_raw.xlsx"
            DBMS=xlsx REPLACE;
     SHEET="LB";
     GETNAMES=YES;
RUN;

/* Begin writing SAS program lb.sas*/

/*show structure of the raw LB data set*/
proc contents data=RAW.LB;
run;
data LB1;

  length USUBJID VISIT $40.;
  set RAW.LB(where=(LBYN="Y") rename=(STUDYID=STUDYID_ LBNAM=LBNAM_
LBNRIND=LBNRIND_ LBFAST=LBFAST_ LBORRES=LBORRES_ LBORNRLO=LBORNRLO_
LBORNRHI=LBORNRHI_ ));
  DOMAIN="LB";
  STUDYID=strip(STUDYID_);
  USUBJID=STRIP(STUDYID)||"-"||STRIP(SITEID)||"-"||STRIP(SUBJID);
  if ^missing(LBDAT) and ^missing(LBTIM) then
LBDTC=put(LBDAT,yymmdd10.)||"T"||put(LBTIM,tod5.);
```

```
  else if ^missing(LBDAT) and missing(LBTIM) then
LBDTC=strip(put(LBDAT,yymmdd10.));
  else if missing(LBDAT) and ^missing(LBTIM) then LBDTC="-----
T"||put(LBTIM,tod5.);
  LBNRIND=upcase(strip(LBNRIND_));
  LBNAM=upcase(strip(LBNAM_));
  LBFAST=upcase(strip(LBFAST_));
  LBORRES=strip(put(LBORRES_,best.));
  LBORRESU=upcase(strip(LBORRESU));
  LBORNRLO=strip(put(LBORNRLO_,best.));
  LBORNRHI=strip(put(LBORNRHI_,best.));

  if index(upcase(EVENT),"SCREENING") then VISIT="SCREENING";
  else if index(upcase(EVENT),"FOLLOW UP") then VISIT="FOLLOW-UP";
  else if index(upcase(EVENT),"DAY") then VISIT="DAY
"||strip(substr(upcase(EVENT),length(EVENT)-1));
  else VISIT=strip(upcase(EVENT));
run;
libname RAW "E://users/directory";

/*PROC IMPORT to import raw EG data in excel to SAS*/
PROC IMPORT OUT= RAW.LB_CONVERTION DATAFILE= "E://users/directory/
LB_conversion.xlsx"
            DBMS=xlsx REPLACE;
    SHEET="CONVERSION";
    GETNAMES=YES;
RUN;

/*pull the LBSTRESU, CONVERTION from RAW.LB_CONVERTION*/

 proc sql noprint;
   create table LB2 as
     select a.*, b.ORIGINAL_UNIT, b.STANDARD_UNIT, b.CONVERSION from LB1 as a
         left join RAW.LB_CONVERTION as b
         on a.LBCAT=b.LBCAT and a.LBTESTCD=b.LBTESTCD and a.LBORRESU=b.
ORIGINAL_UNIT
       order by STUDYID, USUBJID, LBCAT, LBTESTCD, LBDTC
         ;
quit;
data LB3;
  length LBSTRESN  LBSTNRLO LBSTNRHI 8 LBSTRESU LBSTRESC LBTEST LBTESTCD
$40.;
  set LB2(rename=(LBSTRESC=LBSTRESC_ LBSTRESN=LBSTRESN_ LBSTRESU=LBSTRESU_
LBSTNRLO=LBSTNRLO_ LBSTNRHI=LBSTNRHI_
                  LBTESTCD=LBTESTCD_ LBTEST=LBTEST_));
  if substr(LBORRES,1,1)=">" then do;
    LBSTRESN=input(substr(LBORRES,2),best.)*CONVERSION;
    LBSTRESC=">"||strip(put(LBSTRESN,best.));
  end;
  if substr(LBORRES,1,1)="<" then do;
    LBSTRESN=input(substr(LBORRES,2),best.)*CONVERSION;
    LBSTRESC="<"||strip(put(LBSTRESN,best.));
  end;
```

```
  if substr(LBORRES,1,2)=">=" then do;
    LBSTRESN=input(substr(LBORRES,3),best.)*CONVERSION;
    LBSTRESC=">="||strip(put(LBSTRESN,best.));
  end;
  if substr(LBORRES,1,2)="<=" then do;
    LBSTRESN=input(substr(LBORRES,3),best.)*CONVERSION;
    LBSTRESC="<="||strip(put(LBSTRESN,best.));
  end;
  else if ^missing(LBORRES) then do;
    LBSTRESN=input(LBORRES,best.)*CONVERSION;
    LBSTRESC=strip(put(LBSTRESN,best.));
  end;
  LBSTNRHI=input(LBORNRHI,best.)*CONVERSION;
  LBSTNRLO=input(LBORNRHI,best.)*CONVERSION;
  if ^missing(CONVERSION) and ^missing(STANDARD_UNIT) then LBSTRESU=STANDARD_
UNIT;
  LBTEST=upcase(strip(LBTEST_));
  LBTESTCD=upcase(strip(LBTESTCD_));
run;
proc sort data=LB3;
  by USUBJID;
run;
proc sort data=SDTM.EX out=EX(keep=USUBJID EXSTDTC) nodupkey;
  by USUBJID;
run;
data LB4;

/*Derive LBDTC, EXSTDTM*/
  merge LB3(in=a) EX(in=b keep=USUBJID EXSTDTC);
  by USUBJID;
  if length(LBDTC)>=10 then LBDTM=input(substr(LBDTC,1,16),e8601dt.);
  if length(EXSTDTC)>=10 then EXSTDTM=input(substr(EXSTDTC,1,16),e8601dt.);
run;

/* select the condition when LBDTM is on or before EXSTDTM and nonmissing
LBSTRESC, then create the baseline flag "Y" for the last nonmissing result
for each LBTESTCD.*/

/*Sort data set LB4 by USUBJID, LBTESTCD, LBDTC */

proc sort data=LB4 out=BASEFL1;
   by USUBJID LBTESTCD LBDTC;
   where (LBDTM<=EXSTDTM and ^missing(LBSTRESC));
run;
data BASEFL2;
  set BASEFL1;

  /*Derive LBBLFL*/
  by USUBJID LBTESTCD LBDTC ;
  if last.LBTESTCD then LBBLFL="Y";
  keep USUBJID LBTESTCD LBDTC LBBLFL;
run;
proc sort data=BASEFL2;
```

```
   by USUBJID LBTESTCD LBDTC;
run;
proc sort data=LB4;
   by USUBJID LBTESTCD LBDTC;
run;
data LB5;
/*Merge LB4 and BASEFL2*/
   merge LB4(in=a rename=(LBCLSIG=LBCLSIG_)) BASEFL2(in=b);
   by USUBJID LBTESTCD LBDTC;
   if a;
   if LBSTRESC="ABNORMAL" then do;
      if LBCLSIG_="No" then LBCLSIG="N";
      else if LBCLSIG_="Yes" then LBCLSIG="Y";
   end;
 else LBCLSIG="";
run;
/*Sort data set LB5 and SDTM.TV*/
proc sort data=LB5;
   by VISIT;
run;
proc sort data=SDTM.TV out=TV(keep=VISIT VISITNUM) nodupkey;
   by VISIT;
run;
/*Merge LB5 with TV to derive VISITNUM*/
data LB6;
   merge LB5(in=a) TV(in=b);
   by VISIT;
   if a;
run;
proc sort data=LB6;
   by USUBJID LBTESTCD LBDTC VISIT;
run;
/*Derive LBSEQ*/
data Final;
   set LB6;
      by USUBJID LBTESTCD LBDTC VISIT;
      if first.USUBJID then LBSEQ = 0;
      LBSEQ+1;
   output;
   format _all_;
   informat _all_;
run;
libname SDTM "/directory";
data SDTM.LB(label="LABORATORY TEST Results");
/*Assign variable attributes such as label and length to conform with SDTM.LB
Specification (these will also be the same attributes as the SDTM IG).*/

attrib
     STUDYID  label = "Study Identifier"                      length = $20
     DOMAIN   label = "Domain Abbreviation"                   length = $2
```

```
        USUBJID  label = "Unique Subject Identifier"              length = $40
        LBSEQ    label = "Sequence Number"                        length = 8
        LBTESTCD label = "Lab Test or Examination Short Name"      length = $40
        LBTEST   label = "Lab Test or Examination Name"            length = $40
        LBCAT    label = "Category for Lab Test"                   length = $40
        LBORRES  label = "Result or Finding in Original Units"     length = $20
        LBORRESU label = "Original Units"                          length = $40
        LBORNRLO label = "Reference Range Lower Limit in Orig Unit"length = $40
        LBORNRHI label = "Reference Range Upper Limit in Orig Unit"length = $40
        LBSTRESC label = "Character Result/Finding in Std Format"  length = $40
        LBSTRESN label = "Numeric Result/Finding in Standard Units"length = 8
        LBSTRESU label = "Standard Units"                          length = $40
        LBSTNRLO label = "Reference Range Lower Limit-Std Units"   length = 8
        LBSTNRHI label = "Reference Range Upper Limit-Std Units"   length = 8
        LBNRIND  label = "Reference Range for Char Rslt-Std Units" length = $40
        LBBLFL   label = "Baseline Flag"                           length = $2
        LBDTC    label = "Date/Time of Specimen Collection         length = $40
        VISIT    label = "Visit Name"                              length = $40
        VISITNUM label = "Visit Number"                            length = 8
        ;
  set Final;
    keep STUDYID DOMAIN USUBJID LBSEQ LBTESTCD LBTEST LBCAT LBORRES LBORRESU
LBORNRLO LBORNRHI LBSTRESC LBSTRESN LBSTRESU LBSTNRLO LBSTNRHI LBNRIND LBBLFL
LBDTC VISIT VISITNUM
        ;
run;
```

SDTM.TS

```
/*create a libref called RAW pointing to the pathway under E drive*/
libname RAW ".../directory/sdtm_raw";

/*PROC IMPORT to import raw Trial Summary(TS) data in excel to SAS*/
PROC IMPORT OUT= RAW.TS DATAFILE= ".../directory/sdtm_raw.xlsx"
            DBMS=xlsx REPLACE;
     SHEET="TS";
     GETNAMES=YES;
RUN;
data Final;
  set RAW.TS(rename=(STUDYID=STUDYID_ DOMAIN=DOMAIN_ TSSEQ=TSSEQ_  TSPAR-
MCD=TSPARMCD_ TSPARM=TSPARM_ TSVAL=TSVAL_ TSVALNF=TSVALNF_ TSVCDREF=TSVC-
DREF_));

  length DOMAIN $2 TSPARMCD $8 TSSEQ 8.STUDYID $20  TSVCDREF $20
  TSPARM $40 TSVAL TSVALNF  $100;
  STUDYID = strip(STUDYID_);
  DOMAIN  = strip(DOMAIN_);
  TSSEQ=TSSEQ_;
  TSPARMCD=strip(TSPARMCD_);
  TSPARM=strip(TSPARM_);
```

```
    TSVAL=strip(TSVAL_);
    TSVALNF=strip(TSVALNF_);
    TSVCDREF=strip(TSVCDREF_);
    format _all;
    informat _all_;
run;
libname SDTM "E://users/directory";
data SDTM.TS(label="Trial Summary");
  attrib
    STUDYID    label = "Study Identifier"                     length = $20
    DOMAIN     label = "Domain Abbreviation"                  length = $2
    TSSEQ      label = "Sequence Number"                       length = 8
    TSPARMCD   label = "Trial Summary Parameter"              length = $40
    TSPARM     label = "Parameter Value"                       length = $40
    TSVAL      label = "Parameter Value"                       length = $100
    TSVALNF    label = "Parameter Null Flavor"                length = $100
    TSVCDREF   label = "Name of the Reference Terminology"    length = $20
    ;
  set Final;
    keep STUDYID DOMAIN TSSEQ TSPARMCD TSPARM TSVAL TSVALNF TSVCDREF
    ;
run;
```

SDTM.TA

```
/*create a libref called RAW pointing to the pathway under E drive*/
libname RAW ".../directory/sdtm_raw "

/*PROC IMPORT to import raw Trial Arm(TA) data in excel to SAS*/
PROC IMPORT OUT= RAW.TA DATAFILE= ".../directory/sdtm_raw.xlsx"
            DBMS=xlsx REPLACE;
     SHEET="TA";
     GETNAMES=YES;
RUN;

data Final;
  set RAW.TA(rename=(STUDYID=STUDYID_ DOMAIN=DOMAIN_ ARMCD=ARMCD_ ARM=ARM_
TAETORD=TAETORD_
  ETCD=ETCD_ ELEMENT=ELEMENT_ TABRANCH=TABRANCH_ TATRANS=TATRANS_
EPOCH=EPOCH_));
  where DOMAIN_="TA";
  length STUDYID ARMCD $ 20 EPOCH $60. DOMAIN $2. ARM TABRANCH TATRANS $100.
ETCD $8. ELEMENT $100.;
  STUDYID=strip(STUDYID_);
  DOMAIN=strip(DOMAIN_);
  ARMCD=strip(ARMCD_);
  ARM=strip(ARM_);
  TAETORD=TAETORD_;
  ETCD=strip(ETCD_);
  ELEMENT=strip(ELEMENT_);
  TABRANCH=strip(TABRANCH_);
```

```
    TATRANS=TATRANS_;
    EPOCH=strip(EPOCH_);
    format _all_;
    informat _all_;

run;

libname SDTM "E://users/directory";
data SDTM.TA(label="Trial Arm");
  attrib
    STUDYID    label = "Study Identifier"                         length = $20
    DOMAIN     label = "Domain Abbreviation"                      length = $2
    ARM        label = "Description of Planned Arm"               length = $100
    ARMCD      label = "Planned Arm Code"                         length = $20
    TAETORD    label = "Planned Order of Element  within Arm"     length = 8
    ETCD       label = "Element Code"                             length = $8
    ELEMENT    label = "Description of Element"                   length = $100
    TABRANCH   label = "BRANCH"                                   length = $100
    TATRANS    label = "Transition Rule"                          length = $100
    EPOCH      label = "Epoch"                                    length = $60
;
  set Final;
    keep STUDYID DOMAIN ARM ARMCD TAETORD ETCD ELEMENT TABRANCH TATRANS EPOCH
        ;
run;
```

SDTM.TE

```
/*create a libref called RAW pointing to the pathway under E drive*/
libname RAW ".../directory/sdtm_raw";

/*PROC IMPORT to import raw Trial Element(TE) data in excel to SAS*/
PROC IMPORT OUT= RAW.TE DATAFILE= ".../directory/sdtm_raw.xlsx"
            DBMS=xlsx REPLACE;
      SHEET="TE";
      GETNAMES=YES;
RUN;

data Final;
  set RAW.TE(rename=(STUDYID=STUDYID_ DOMAIN=DOMAIN_ ETCD=ETCD_
ELEMENT=ELEMENT_ TESTRL=TESTRL_ TEENRL=TEENRL_ ));
  where DOMAIN_="TE";
  length STUDYID $20. DOMAIN $2. ETCD $8. TESTRL TEENRL ELEMENT $100.;
  STUDYID=STUDYID;
  DOMAIN=DOMAIN;
  ETCD=ETCD;
  ELEMENT=ELEMENT;
  TESTRL=TESTRL;
  TEENRL=TEENRL;
  format _all;
  informat _all;
```

```
run;

libname SDTM ".../directory/sdtm_raw";
data SDTM.TE(label="Trial Element");
  attrib
      STUDYIDlabel = "Study Identifier"                       length = $20
      DOMAIN label = "Domain Abbreviation"                    length = $2
      ETCD        label = "Element Code"                       length = $8
      ELEMENTlabel = "Description of Element"                   length = $100
      TESTRL label = "Rule for Start of Element"               length = $100
      TEENRL label = "Rule for End of Element"                 length = $100
;
  set Final;
    keep STUDYID DOMAIN ETCD ELEMENT TESTRL TEENRL
        ;
run;
```

SDTM.TV

```
/*create a libref called RAW pointing to the pathway under E drive*/
libname RAW ".../directory/sdtm_raw";

/*PROC IMPORT to import raw Trial Visit(TV) data in excel to SAS*/
PROC IMPORT OUT= RAW.TV DATAFILE= ".../directory/sdtm_raw.xlsx"
    DBMS=xlsx REPLACE;
    SHEET="TV";
    GETNAMES=YES;
RUN;

data Final;
  set RAW.TV(rename=(STUDYID=STUDYID_ DOMAIN=DOMAIN_ VISITNUM=VISITNUM_
VISIT=VISIT_
  ARM=ARM_ ARMCD=ARMCD_ TVSTRL=TVSTRL_ ));
  length STUDYID ARMCD $20. DOMAIN $2. ARM TVSTRL $100. VISIT $40.;
  STUDYID=strip(STUDYID_);
  DOMAIN=strip(DOMAIN_);
  VISITNUM=VISITNUM_;
  VISIT=strip(VISIT_);
  ARM=strip(ARM_);
  ARMCD=strip(ARMCD_);
  TVSTRL=strip(TVSTRL_);

  format _all_;
  informat _all_;
run;

libname SDTM ".../directory/sdtm_raw"
data SDTM.TV(label="Trial Visit");
   attrib
      STUDYID        label = "Study Identifier"                length = $20
      DOMAIN         label = "Domain Abbreviation"             length = $2
      VISITNUM       label = "Visit Number"                     length = 8
```

```
            VISIT          label = "Visit Name"                              length = $40
            ARM            label = "Planned Arm"                             length = $100
            ARMCD          label = "Planned Arm Code"                        length = $20
            TVSTRL         label = "Visit Start Rule"                        length = $100
;
  set Final;
    keep STUDYID DOMAIN ARM ARMCD VISIT VISITNUM TVSTRL
        ;
run;
```

SDTM.TI

```
libname RAW ".../directory/sdtm_raw";
/*PROC IMPORT to import raw Trial Arm(TA) data in excel to SAS*/
PROC IMPORT OUT= RAW.TI DATAFILE= ".../directory/sdtm_raw.xlsx"
     DBMS=xlsx REPLACE;
     SHEET="TI";
     GETNAMES=YES;
RUN;

data Final;
  set RAW.TI(rename=(STUDYID=STUDYID_ DOMAIN=DOMAIN_ IETESTCD=IETESTCD_
IETEST=IETEST_ IECAT=IECAT_));
  where DOMAIN_="TI";
  length STUDYID $20. DOMAIN $2. IETESTCD IETEST $100. IECAT $40.;
  STUDYID=strip(STUDYID_);
  DOMAIN=strip(DOMAIN_);
  IETESTCD=strip(IETESTCD_);
  IETEST=strip(IETEST_);
  IECAT=strip(IECAT_);

  format _all_;
  informat _all_;
run;

libname SDTM ".../directory/sdtm_raw";
data SDTM.TI(label="Trial Inclusion/Exclusion Criteria");
  attrib
    STUDYID    label = "Study Identifier"                              length = $20
    DOMAIN     label = "Domain Abbreviation"                           length = $2
    IETESTCD   label = "Inclusion/Exclusion Criterion Short Name" length = $100
    IETEST     label = "Inclusion/Exclusion Criterion"                length = $100
    IECAT      label = "Inclusion/Exclusion Category"                 length = $40
;
  set Final;
    keep STUDYID DOMAIN IETESTCD IETEST IECAT
;
run;
```

A.3 ADaM Programming Section
ADaM.ADSL

```
/*Begin writing SAS program sdtm.dm.sas*/
/*Demographic Variables*/
data dm1;
   set sdtm.dm;
   length AETHNIC $40.;
   if missing(ETHNIC) then AETHNIC = "NOT COLLECTED";
   else AETHNIC = ETHNIC;
   If RACE = "WHITE" then ARACE = "W";
   else if RACE = "BLACK OR AFRICAN AMERICAN" then ARACE = "B";
   else if RACE = "NATIVE HAWAIIAN OR OTHER PACIFIC ISLANDERS" then ARACE =
"HP";
   else if RACE = "ASIAN" then ARACE = "A";
   else if RACE = "AMERICAN INDIAN OR ALASKA AMERICAN" then ARACE = "AA";
/*Derive TRTSDT, TRTSTM, TRTEDT, TRTETM */
  TRTSDT = input(substr(RFSTDTC,1,10),yymmdd10.);
  TRTSTM = input(substr(RFSTDTC,12),time5.);
  TRTEDT = input(substr(RFENDTC,1,10),yymmdd10.);
  TRTETM = input(substr(RFENDTC,12),time5.);
  format TRTSDT yymmdd10. TRTSTM time5. TRTEDT yymmdd10. TRTETM time5.;
  length TRT01P TRT01A $20.;
/*Derive TRT01P, TRT01A*/
  if ARMCD = "DRUG A" then TRT01P = "TRTA";
  if ACTARMCD = "DRUG A" then TRT01A = "TRTA";
 /*Derive SAFFL, FASFL */
  if ^missing(RFSTDTC) then do;
    SAFFL="Y";
    FASFL="Y";
  end;
  else do;
  SAFFL="N";
  FASFL="N";
  end;
run;

/*Merge dm2 and dm1 to catch DSDECOD, DSSTDTC*/
proc sql;
   create table dm2 as select a.*,b.DSDECOD,DSSTDTC from dm1 a left join sdtm.
ds b on a.USUBJID=b.USUBJID;
quit;
data dm3;
  set dm2;
  length EOSDT 8 EOSSTT $20.;
  EOSDT = input(substr(DSSTDTC,1,10),yymmdd10.);
  if DSDECOD = "COMPLETED" then EOSSTT = "COMPLETED";
  else EOSSTT = "DISCONTINUED";
```

```
   format EOSDT yymmdd10.;
run;

/*Sort data set SDTM.EX by USUBJID, EXSTDTC without duplicate values*/
proc sort data=SDTM.EX out=ex nodupkey;
   by USUBJID EXSTDTC EXENDTC;
run;
data ex1
     ex2;
  set ex;
  by USUBJID EXSTDTC EXENDTC;

  /*Create data set ex1 if the first.USUBJID statement*/
/*Create data set ex2 if the last.USUBJID statement*/

  if First.USUBJID then output ex1;
  if Last.USUBJID then output ex2;
run;
proc sql;
/*Create data set dm4 and dm5*/

  create table dm4 as select a.*, b.EXSTDTC from dm3 a left join ex1 b on
a.USUBJID=b.USUBJID;

  create table dm5 as select a.*, b.EXENDTC from dm4 a left join ex2 b on
a.USUBJID=b.USUBJID;
quit;
data Final;
  set dm5;

  /*Derive TRT01SDTM, TRT01SDT, TRT01EDTM, TRT01EDT */
  length TRT01SDTM TRT01SDT TRT01EDTM TRT01EDT 8.;
  TRT01SDTM = input(EXSTDTC, e8601dt.);
  TRT01SDT = input(substr(EXSTDTC,1,10),yymmdd10.);
  TRT01EDTM = input(EXENDTC, e8601dt.);
  TRT01EDT = input(substr(EXENDTC,1,10),yymmdd10.);
  format TRT01SDTM TRT01EDTM e8601dt. TRT01SDT TRT01EDT yymmdd10.;
run;

libname ADAM ".../directory";
data ADAM.ADSL(label="Subject-Level Analysis Data Set");

/*Assign variable attributes such as label and length to conform with
ADAM.ADSL Specification (these will also be the same attributes as the ADAM
IG).*/

attrib
 STUDYID        label = "Study Identifier"                    length = $20
 USUBJID        label = "Unique Subject Identifier"           length = $40
 SUBJID         label = "Subject Identifier for the Study"    length = $20
 SITEID         label = "Study Site Identifier"               length = $10
 BRTHDTC        label = "Date/Time of Brith"                  length = $20
 AGE            label = "Age"                                 length = 8
 AGEU           label = "Age Units"                           length = $10
```

```
   SEX            label = "Sex"                                 length = $2
   RACE           label = "Race"                                length = $100
   ARACE          label = "Analysis Race"                       length = $100
   ETHNIC         label = "Ethnicity"                           length = $60
   AETHNIC        label = "Analysis Ethnicity"                  length = $60
   SAFFL          label = "Safety Population Flag"              length = $1
   FASFL          label = "Full Analysis Set Population Flag"   length = $1
   ARM            label = "Description of Planned Arm"          length = $200
   ARMCD          label = "Planned Arm Code"                    length = $20
   ACTARMCD       label = "Actual Arm Code"                     length = $20
   ACTARM         label = "Description of Actual Arm"           length = $200
   TRT01P         label = "Planned Treatment for Period 1"      length = $20
   TRT01A         label = "Actual Treatment for Period 1"       length = $20
   RFSTDTC        label = "Subject Reference Start Date/Time"   length = $20
   RFENDTC        label = "Subject Reference End Date/Time"     length = $20
   TRTSDT         label = "Date of First Exposure to Treatment" length = 8
   TRTSTM         label = "Time of First Exposure to Treatment" length = 8
   TRT01SDTM      label = "Datetime of First Exposure in Period 1" length = 8
   TRT01SDT       label = "Date of First Exposure in Period 1"  length = 8
   TRTEDT         label = "Date of Last Exposure to Treatment"  length = 8
   TRTETM         label = "Time of Last Exposure to Treatment"  length = 8
   TRT01EDTM      label = "Datetime of Last Exposure in Period 1" length = 8
   TRT01EDT       label = "Date of Last Exposure in Period 1"   length = 8
   EOSSTT         label = "End of Study Status"                 length = $20
   EOSDT          label = "End of Study"                        length = 8
   COUNTRY        label = "Country"                             length = $4
   ;
set Final;
  keep STUDYID USUBJID SUBJID SITEID BRTHDTC AGE AGEU SEX RACE ARACE ETHNIC
  AETHNIC SAFFL FASFL ARM ARMCD ACTARMCD ACTARM TRT01P TRT01A RFSTDTC RFENDTC
  TRTSDT TRTSTM TRT01SDTM TRT01SDT TRTEDT TRTETM TRT01EDTM TRT01EDT EOSSTT
  EOSDT COUNTRY
       ;
run;
```

ADaM.ADEG

```
/*Begin writing SAS program merge SDTM.EG and ADAM.ADSL.sas*/
data ADEG1;
  merge SDTM.EG(in=a) ADAM.ADSL(in=b drop=STUDYID);
  by USUBJID;
  if a and b;
run;

/*Derive PARAM*/
proc sql;
    create table ADEG2 as select distinct EGTEST,EGORRESU from ADEG1 where
EGORRESU ne '';
quit;

data ADEG3;
```

```
  length PARAM  $40;
  set ADEG2;
  PARAM=strip(EGTEST)||" ("||strip(EGORRESU)||")";
  keep EGTEST PARAM;
run;
proc sql;
/*Left Join PARAM from ADEG1 with ADEG3 when the same EGTEST*/
    create table ADEG4 as select a.*,b.PARAM from ADEG1 as a left join ADEG3
as b on a.EGTEST=b.EGTEST;

      /*Map USUBJID, EGTESTCD, and the number of USUBJID from ADEG4 to
ADEG5*/
    create table ADEG5 as select USUBJID,EGTESTCD,count(USUBJID) as count
from ADEG4
     group by USUBJID,EGTESTCD;

      /*Left Join ADEG4 with ADEG5 when the USUBJID and EGTESTCD are the
same */
    create table ADEG6 as select a.*,b.count from ADEG4 as a left join ADEG5
as b
    on a.USUBJID=b.USUBJID and a.EGTESTCD=b.EGTESTCD;
quit;

/*Derive DTYPE*/
data ADEG7;
  set ADEG6(where=(EGTESTCD^="INTP"));
  length DTYPE $20;
  EGSTRESN=input(EGSTRESC,best.);
  by USUBJID EGTESTCD;
  if first.EGTESTCD then do;
     SUM=EGSTRESN;
     N=1;
  end;
  else do;
     SUM+EGSTRESN;
     N+1;
  end;
  output;
  if last.EGTESTCD then do;
     DTYPE="AVERAGE";
     FLAG=1;
  output;
end;
run;

/*Derive EGSTRESN and EGSTRESC*/

data ADEG8;
  set ADEG7;
  if DTYPE="AVERAGE" then do;
    if SUM ne . then do;
     EGSTRESN_MEAN=SUM/N;
     EGSTRESC=strip(put(EGSTRESN,best.));
```

```
    end;
  else  do;
    EGSTRESN=.;
    EGSTRESC="";
    end;
 end;
run;
/*Set ADEG8 and ADEG6 with EGTESTCD="INTP"*/
data ADEG9;
  set ADEG8 ADEG6(where=(EGTESTCD="INTP"));
run;
/*Sort data set ADEG9 by USUBJID, EGTESTCD, EGDTC */
proc sort data=ADEG9;
      by USUBJID EGTESTCD EGDTC;
run;
data ADEG10;
  set ADEG9;
 /*Derive ADT, ATM, ADTM */
  length PARAMCD $8. AVALC $40.;
  if length(EGDTC)=10 then do;
  ADT=input(EGDTC,yymmdd10.);
  ATM=.;
  ADTM=.;
  end;
   if length(EGDTC)>10 then do;
  ADTM=input(EGDTC,is8601dt.);
  ADT=datepart(ADTM);
  ATM=timepart(ADTM);
  end;
  format ADTM is8601dt. ADT yymmdd10. ATM time5.;
 /*Derive ADY */
  if nmiss(ADT,TRTSDT)=0 then ADY=ADT-TRTSDT+(ADT>=TRTSDT);
   /*Derive APHASE,EMDESC*/
  if (ADT<=TRTSDT and ATM=.) or (ADT^=. and ATM^=. and ADTM<=TRT01SDTM) then
do;
    APHASE="Screening";
    EMDESC="P";
  end;
  if (ADT > TRTSDT and ATM =.) or (ADT^=. and ATM^=. and ADTM>TRT01SDTM) then
do;
    APHASE="Treatment";
    EMDESC="T";
  end;
  if (ADT > TRTEDT and ATM =.) or (ADT^=. and ATM^=. and ADTM>TRT01EDTM) then
do ;
    APHASE="Follow-Up";
    EMDESC="A";
  end;
```

```
 /*Derive PARAMCD */
  if EGTEST='Interpretation' then PARAM=strip(EGTEST);
  else PARAM=PARAM;
  PARAMCD=strip(EGTESTCD);
/*Derive AVAL and AVALC */
  if PARAMCD^='INTP' and DTYPE='AVERAGE' then AVAL=EGSTRESN_MEAN;
  else if PARAMCD^='INTP' and DTYPE='' then AVAL=EGSTRESN;
  else if PARAMCD='INTP' then AVAL=.;

  if PARAMCD='INTP' then AVALC=strip(EGSTRESC);
  else if PARAMCD^='INTP' then AVALC='';
/*Derive TRTP, TRTA */
  TRTP=TRT01P;
  TRTA=TRT01A;

run;

/*Derive ABLFL */
data ADEG10;
  set ADEG10;
  NUMBER=_n_;
run;

/*Sort data set ADEG10 by USUBJID, PARAMCD, ADT, ADTM, DTYPE */
proc sort data=ADEG10;
  by USUBJID PARAMCD ADT ADTM DTYPE;
run;

/*Filter the condition of baseline flag */
data BASE;
  set ADEG10(where=(EMDESC="P" and (AVAL ne . or AVALC ne '')  and
(.<ADT<=TRTSDT) and COUNT>1));
  by USUBJID PARAMCD ADT ADTM DTYPE;
run;

/*if the last PARAMCD then ABLFL sets to "Y" */
data ABLFL;
  set BASE;
  by USUBJID PARAMCD ADT ADTM DTYPE;
  if last.PARAMCD; ABLFL="Y";
run;

/*Left Join ADEG10 with ABLFL */
proc sql;
      create table ADEG11 as select a.*,b.ABLFL from ADEG10 as a left join
ABLFL as b on a.NUMBER=b.NUMBER;
quit;

/*Derive variable Base */
proc sql;
      create table ADEG12 as select a.*,b.AVAL as BASE from ADEG11 as a
    left join ADEG11(where=(ABLFL='Y')) as b on a.USUBJID=b.USUBJID and
a.PARAMCD=b.PARAMCD;
quit;
```

```
/*Derive variable CHG*/
data ADEG13;
  set ADEG12;
  if n(AVAL,BASE)=2 and ABLFL ne "Y"  then CHG=AVAL-BASE;
  if ABLFL^="Y" and EMDESC="P" then do;
  CHG=.;
 end;
run;

/*Derive variable AVISIT and AVISITN*/

data ADEG14;
  set ADEG13;
  length AVISIT $40. AVISITN 8. ;

  if PARAMCD ne "INTP" then do;
   if ABLFL="Y" then do;
      AVISIT="Baseline";
      AVISITN=0;
    end;

   else if index(VISIT,"FOLLOW-UP") then do;
      AVISIT="Follow-up";
      AVISITN=100;
   end;
   else do;
      AVISIT=strip(VISIT);
      AVISITN=input(compress(AVISIT,,"kd"),best.);
        end;
   end;
run;

/*Sort data set ADEG14*/
proc sort data=ADEG14; by USUBJID PARAMCD ADT ATM DTYPE EGSEQ;run;

/*Derive ASEQ*/
data Final;
   set ADEG14;
     by USUBJID PARAMCD ADT ATM DTYPE EGSEQ;
     if first.USUBJID then ASEQ = 0;
     ASEQ+1;
  output;
run;
libname ADAM ".../directory";
data ADAM.ADEG(label="ECG Test Results Analysis Data sets");

/*Assign variable attributes such as label and length to conform with
ADAM.ADSL Specification (these will also be the same attributes as the ADAM
IG).*/

attrib
  STUDYID      label = "Study Identifier"               length = $20
  USUBJID      label = "Unique Subject Identifier"      length = $40
```

```
    SUBJID       label = "Subject Identifier for the Study"         length = $20
    EGSEQ        label = "Sequence Number"                          length = 8
    ASEQ         label = "Analysis Sequence Number"                 length = 8
    TRTP         label = "Planned Treatment"                        length = $40
    TRTA         label = "Actual Treatment"                         length = $40
    ADT          label = "Analysis Date"                            length = 8
    ATM          label = "Analysis Time"                            length = 8
    ADTM         label = "Analysis Date and Time"                   length = 8
    ADY          label = "Analysis Relative Day"                    length = 8
    AVISIT       label = "Analysis Visit"                           length = $40
    AVISITN      label = "Analysis Visit (N)"                       length = 8
    APHASE       label = "PHASE"                                    length = $40
    PARAM        label = "Parameter"                                length = $40
    PARAMCD      label = "Parameter Code"                           length = $8
    AVAL         label = "Analysis Value"                           length = 8
    AVALC        label = "Analysis Value (C)"                       length = $40
    ABLFL        label = "Baseline Record Flag"                     length = $1
    BASE         label = "Baseline Value"                           length = 8
    CHG          label = "Change from Baseline"                     length = 8
    DTYPE        label = "Derivation Type"                          length = $20
    EMDESC       label = "Description of Treatment Emergent"        length = $20
    EGORRES      label = "Result of Finding in Original Units"      length = $100
    EGORRESU     label = "Original Units"                           length = $40
    EGSTRESC     label = "Character Result/ Finding in Std Format"  length = $100
    VISIT        label = "Visit Name"                               length = $40
    VISITNUM     label = "Visit Number"                             length = 8
    EGDTC        label = "Date/Time of ECG"                         length = $40
  ;
  set Final;
    keep STUDYID USUBJID SUBJID EGSEQ ASEQ TRTP TRTA ADT ATM ADTM ADY AVISIT
    AVISITN APHASE PARAM PARAMCD AVAL AVALC ABLFL BASE CHG DTYPE EMDESC
    EGORRES EGORRESU EGSTRESC VISIT VISITNUM EGDTC
        ;
run;
```

ADaM.ADLB

```
/*Begin writing SAS program merge SDTM.LB and ADAM.ADSL.sas*/
data ADLB1;
  merge SDTM.LB(in=a) ADAM.ADSL(in=b drop=STUDYID);
  by USUBJID;
  if a and b;
run;
data ADLB2;
  set ADLB1;

/*Derive ADT, ATM, ADTM*/
  length PARAMCD $8. PARAM AVALC $40. AVAL 8;
      if length(LBDTC)=10 then do;
```

```
        ADT=input(LBDTC,yymmdd10.);
        ATM=.;
        ADTM=.;
end;

if length(LBDTC)>10 then do;
        ADTM=input(LBDTC,is8601dt.);
        ADT=datepart(ADTM);
        ATM=timepart(ADTM);
  end;
  format ADTM is8601dt. ADT yymmdd10. ATM time5.;

  /*Derive ADY*/
  if nmiss(ADT,TRTSDT)=0 then ADY=ADT-TRTSDT+(ADT>=TRTSDT);
/*Derive APHASE, EMDESC*/

  if (ADT<=TRTSDT and ATM=.) or (ADT^=. and ATM^=. and ADTM<=TRT01SDTM) then
do;
    APHASE="Screening";
    EMDESC="P";
  end;
  if (ADT > TRTSDT and ATM =.) or (ADT^=. and ATM^=. and ADTM>TRT01SDTM) then
do;
    APHASE="Treatment";
    EMDESC="T";
  end;
  if (ADT > TRTEDT and ATM =.) or (ADT^=. and ATM^=. and ADTM>TRT01EDTM) then
do ;
    APHASE="Follow-Up";
    EMDESC="A";
  end;


/*Derive PARAM, PARAMCD*/
  PARAM=strip(LBTEST)||" ("||strip(LBORRESU)||")";
  PARAMCD=strip(LBTESTCD);

/*Derive AVAL, AVALC, DTYPE*/

  if ^missing(LBSTRESN) then do;
    AVAL=LBSTRESN;
    DTYPE="";
  end;
else if LBSTRESN=. and ^missing(LBSTRESC) then do;
   if index(LBSTRESC,"<") or index(LBSTRESC,"<=") then do;
       AVAL=input(compress(LBSTRESC,"<="),best.);
       DTYPE="IMPUTE";
   end;
   if index(LBSTRESC,">") or index(LBSTRESC,">=") then do;
       AVAL=input(compress(LBSTRESC,">="),best.);
       DTYPE="IMPUTE";
   end;
  end;
```

```
  AVALC=strip(LBSTRESC);

/*Derive TRTP, TRTA*/
  TRTP=TRT01P;
  TRTA=TRT01A;
run;

/*Derive ABLFL*/
data ADLB2;
  set ADLB2;
  NUMBER=_n_;
run;


proc sort data=ADLB2;
  by USUBJID PARAMCD ADT ADTM;
run;

/*Filter the condition of baseline flag */
data BASE;
  set ADLB2(where=(EMDESC="P" and (AVAL ne . or AVALC ne '')  and
(.<ADT<=TRTSDT)));
  by USUBJID PARAMCD ADT ADTM;
run;

 /*if the last PARAMCD then ABLFL sets to "Y" */
data ABLFL;
      set BASE;
      by USUBJID PARAMCD ADT ADTM;
      if last.PARAMCD; ABLFL="Y";
run;

/*Left Join ADLB2 with ABLFL */
proc sql;
      create table ADLB3 as select a.*,b.ABLFL from ADLB2 as a left join
ABLFL as b on a.NUMBER=b.NUMBER;
quit;

/*Derive variable Base */
proc sql;
      create table ADLB4 as select a.*,b.AVAL as BASE from ADLB3 as a
    left join ADLB3(where=(ABLFL='Y')) as b on a.USUBJID=b.USUBJID and
a.PARAMCD=b.PARAMCD;
quit;
/*Derive variable CHG*/
data ADLB5;
      set ADLB4;
    if n(AVAL,BASE)=2 and ABLFL ne "Y"  then CHG=AVAL-BASE;
      if ABLFL^="Y" and EMDESC="P" then do; CHG=.;end;
run;
```

```
/*Derive variable AVISIT and AVISITN*/

data ADLB6;
  set ADLB5;
  length AVISIT $40. AVISITN 8. ;
    if ABLFL="Y" then do;
       AVISIT="Baseline";
       AVISITN=0;
    end;
     else if index(VISIT,"FOLLOW-UP") then do;
      AVISIT="Follow-up";
      AVISITN=100;
     end;
    else do;
      AVISIT=strip(VISIT);
      AVISITN=input(compress(AVISIT,,"kd"),best.);
      end;
run;
proc sort data=ADLB6; by USUBJID PARAMCD ADT ATM DTYPE LBSEQ;run;

/*Derive ASEQ*/
data Final;
  set ADLB6;
   by USUBJID PARAMCD ADT ATM DTYPE LBSEQ;
   if first.USUBJID then ASEQ = 0;
    ASEQ+1;
  output;
run;


libname ADAM "..././directory";
data ADAM.ADLB(label="Lab Test Results Analysis Data sets");

/*Assign variable attributes such as label and length to conform with
ADAM.ADSL Specification (these will also be the same attributes as the ADAM
IG).*/

attrib
STUDYID         label = "Study Identifier"                    length = $20
USUBJID         label = "Unique Subject Identifier"           length = $40
SUBJID          label = "Subject Identifier for the Study"    length = $20
LBSEQ           label = "Sequence Number"                     length = 8
ASEQ            label = "Analysis Sequence Number"            length = 8
TRTP            label = "Planned Treatment"                   length = $40
TRTA            label = "Actual Treatment"                    length = $40
ADT             label = "Analysis Date"                       length = 8
ATM             label = "Analysis Time"                       length = 8
ADTM            label = "Analysis Date and Time"              length = 8
ADY             label = "Analysis Relative Day"               length = 8
AVISIT          label = "Analysis Visit"                      length = $40
AVISITN         label = "Analysis Visit (N)"                  length = 8
APHASE          label = "PHASE"                               length = $40
```

```
PARAM          label = "Parameter"                                   length = $40
PARAMCD        label = "Parameter Code"                              length = $8
AVAL           label = "Analysis Value"                              length = 8
AVALC          label = "Analysis Value (C)"                          length = $40
ABLFL          label = "Baseline Record Flag"                        length = $1
BASE           label = "Baseline Value"                              length = 8
CHG            label = "Change from Baseline"                        length = 8
DTYPE          label = "Derivation Type"                             length = $20
EMDESC         label = "Description of Treatment Emergent"           length = $20
LBORRES        label = "Result of Finding in Original Units"         length = $100
LBORRESU       label = "Original Units"                              length = $40
LBORNRLO       label = "Reference Range Lower Limit in Orig Unit"    length = $40
LBORNRHI       label = "Reference Range Higher Limit in Orig Unit"   length = $40
LBSTNRLO       label = "Reference Range Lower Limit-Std Units"       length = 8
LBSTNRHI       label = "Reference Range Upper Limit-Std Units"       length = 8
LBSTRESC       label = "Character Result/ Finding in Std Format"     length = $100
VISIT          label = "Visit Name"                                  length = $40
VISITNUM       label = "Visit Number"                                length = 8
LBDTC          label = "Date/Time of Lab"                            length = $40
      ;
  set Final;
    keep STUDYID USUBJID SUBJID LBSEQ ASEQ TRTP TRTA ADT ATM ADTM ADY AVISIT
AVISITN APHASE PARAM PARAMCD AVAL AVALC ABLFL BASE CHG DTYPE EMDESC LBORRES
LBORRESU LBORNRLO LBSTNRHI LBSTNRLO LBSTNRHI LBSTRESC VISIT VISITNUM
LBDTC
      ;
run;
```

ADaM.ADAE

```
/*Begin writing SAS program merge SDTM.AE and ADAM.ADSL.sas*/
data ADAE1;
  merge SDTM.AE(in=a) ADAM.ADSL(in=b drop=STUDYID);
  by USUBJID;
  if a and b;
run;
data FINAL;
  set ADAE1;
  length TRTEMFL PREFL ASTDTF $1 ASTDT AENDT 8. ;

  /*Derive ASTDT, AENDT */
  ASTDT=input(substr(AESTDTC,1,10),yymmdd10.);
  AENDT=input(substr(AEENDTC,1,10),yymmdd10.);
  format ASTDT AENDT yymmdd10.;

  /*Derive TRTEMFL, PREFL, FUPFL, EMDESC*/
  if TRTSDT <= ASTDT <= TRTEDT then do;
    TRTEMFL = "Y";
    EMDESC = "T";
  end;
  if . < ASTDT < TRTSDT then do;
```

```
      PREFL = "Y";
      EMDESC = "P";
  end;
   else if ASTDT > TRTEDT and TRTEDT^=. then do;
     FUPFL = "Y";
     EMDESC= "A";
   end;
   if nmiss(ASTDT,TRTSDT)=0 then ASTDY=ASTDT-TRTSDT+(ASTDT>=TRTSDT);

  /*Derive ASTDTF*/

   if length(AESTDTC) >= 10 then ASTDTF="";
   else if length(AESTDTC)=7 then ASTDTF="D";
   else if length(AESTDTC)=4 then ASTDTF="M";

    /*Derive TRTP, TRTA*/
   TRTA=TRT01A;
   TRTP=TRT01P;
  run;
libname ADAM ".../directory";
data ADAM.ADAE(label="Adverse Events Analysis Data Set");

/*Assign variable attributes such as label and length to conform with ADAM.
ADSL Specification (these will also be the same attributes as the ADAM IG).*/


  attrib
   STUDYID       label = "Study Identifier"                         length = $20
   USUBJID       label = "Unique Subject Identifier"                length = $40
   AESEQ         label = "Sequence Number"                          length = 8
   AETERM        label = "Reported Term for the Adverse Event"      length = $200
   AELLT         label = "Lowest Level Term"                        length = $100
   AELLTCD       label = "Lowest Level Term Code"                   length = 8
   AEDECOD       label = "Dictionary-Derived Term"                  length = $200
   AEPTCD        label = "Preferred Term Code"                      length = 8
   ASTDT         label = "Analysis Start Date"                      length = 8
   ASTDTF        label = "Analysis Start Date Imputation Flag"      length = $1
   AEENDTC       label = "End Date/Time of Adverse Event"           length = $20
   AENDT         label = "Analysis Start Date"                      length = 8
   ASTDY         label = "Analysis Start Relative Day"              length = 8
   TRTEMFL       label = "Treatment Emergent Analysis Flag"         length = $1
   PREFL         label = "Pre-treatment Flag"                       length = $1
   FUPFL         label = "Follow-Up Flag"                           length = $1
   EMDESC        label = "Description of Treatment Emergent "       length = $20
   AEHLT         label = "High-Level Term"                          length = $200
   AEHLTCD       label = "High-Level Term Code"                     length = 8
   AEHLGT        label = "High-Level Group Term"                    length = $200
   AEHLGTCD      label = "High-Level Group Term Code"               length = 8
   AEBODSYS      label = "Body System or Organ Class"               length = $20
   AESER         label = "Serious Event"                            length = $2
   AEACN         label = "Action Taken with Study Treatment"        length = $50
   AEREL         label = "Causality"                                length = $50
   AEOUT         label = "Outcome of Adverse Event"                 length = $50
```

```
    AESCONG        label = "Congenital Anomaly or Birth Defect"        length = $2
    AESDISAB       label = "Persist or Significant Disability"         length = $2
    AESDTH         label = "Results in Death"                          length = $2
;
    set FINAL;

    keep STUDYID USUBJID AESEQ AETERM AELLT AELLTCD AEDECOD AEPTCD ASTDT
ASTDTF AEENDTC AENDT ASTDY TRTEMFL PREFL FUPFL EMDESC AEHLT AEHLTCD AEHLGT
AEHLGTCD AEBODSYS AESER AEACN AEREL AEOUT AESCONG AESDISAB AESDTH
        ;
run;
```

ADaM.ADCM

```
/*Begin writing SAS program merge SDTM.CM and ADAM.ADSL.sas*/

data ADCM1;
  merge SDTM.CM(in=a) ADAM.ADSL(in=b drop=STUDYID);
  by USUBJID;
  if a and b;
run;
data ADCM2;
  set ADCM1;
  length ADURU $20 ONTRTFL PREFL ASTDTF $1 ASTDT AENDT ADURN 8. ;

  /*Derive ASTDT, AENDT */
  ASTDT=input(substr(CMSTDTC,1,10),yymmdd10.);
  AENDT=input(substr(CMENDTC,1,10),yymmdd10.);
  format ASTDT AENDT yymmdd10.;
 /*Derive ONTRTFL, PREFL, FUPFL, EMDESC*/

  if TRTSDT <= ASTDT <= TRTEDT then ONTRTFL = "Y";
  else if . < ASTDT < TRTSDT then PREFL = "Y";
  else if ASTDT > TRTEDT and TRTEDT^=. then  FUPFL = "Y";
/*Derive ASTDY*/

  if nmiss(ASTDT,TRTSDT)=0 then ASTDY=ASTDT-TRTSDT+(ASTDT>=TRTSDT);
/*Derive ASTDTF*/

  if length(CMSTDTC) >= 10 then ASTDTF="";
  else if length(CMSTDTC)=7 then ASTDTF="D";
  else if length(CMSTDTC)=4 then ASTDTF="M";
/*Derive ADURN, ADURU*/

  if cmiss(CMSTDTC,CMENDTC)=0 then ADURN=AENDT-ASTDT+1;
  if ADURN>1 then ADURU="DAYS";
  else if ADURN=1 then ADURU="DAY";
  else ADURU="";
/*Derive TRTP, TRTA*/

  TRTA=TRT01A;
```

```
  TRTP=TRT01P;
 run;
 proc sort data=ADCM2;
   by USUBJID CMTRT ASTDT AENDT CMDECOD CMSEQ;
run;

 /*Derive ASEQ*/

data Final;
  set ADCM2;
  by USUBJID CMTRT ASTDT AENDT CMDECOD CMSEQ;
  if first.USUBJID then ASEQ = 0;
   ASEQ+1;
  output;
run;

libname ADAM "..../directory";

data ADAM.ADCM(label="Concomitant Medications Analysis Data Set");

/*Assign variable attributes such as label and length to conform with
ADAM.ADSL Specification (these will also be the same attributes as the ADAM
IG).*/

attrib
STUDYID    label = "Study Identifier"                          length = $20
USUBJID    label = "Unique Subject Identifier"                 length = $40
CMSEQ      label = "Sequence Number"                           length = 8
ASEQ       label = "Analysis Sequence Number"                  length = 8
CMDECOD    label = "Dictionary-Derived Term"                   length = $200
ASTDT      label = "Analysis Start Date"                       length = 8
ASTDTF     label = "Analysis Start Date Imputation Flag"       length = $1
ONTRTFL    label = "On Treatment Record Flag "                 length = $1
PREFL      label = "Pre-treatment Flag"                        length = $1
FUPFL      label = "Follow-Up Flag"                            length = $1
AENDT      label = "Analysis Start Date"                       length = 8
ASTDY      label = "Analysis Start Relative Day"               length = 8
CMTRT      label = "Reported name of drug, Medication or Therapy" length = $200
CMINDC     label = "Indication"                                length = $200
CMROUTE    label = "Route of Administration"                   length = $40
CMSTDTC    label = "Start Date/Time of Medication"             length = $20
CMENDTC    label = "End Date/Time of Medication"               length = $20
CMENRTPT   label = "End Relative to Reference Time Point"       length = $20
      ;
   set FINAL;
   keep STUDYID USUBJID CMSEQ CMDECOD ASTDT ASTDTF ONTRTFL PREFL FUPFL AENDT
ASTDY CMTRT CMINDC CMROUTE CMSTDTC CMENDTC CMENRTPT
;
run;
```

Ready to take your SAS® and JMP® skills up a notch?

Be among the first to know about new books,
special events, and exclusive discounts.
support.sas.com/newbooks

Share your expertise. Write a book with SAS.
support.sas.com/publish

Continue your skills development with free online learning.
https://www.sas.com/en_us/training/offers/free-training.html

 sas.com/books
for additional books and resources.

CPSIA information can be obtained
at www.ICGtesting.com
Printed in the USA
BVHW021442170223
658736BV00007B/758